The Development of New Food Products

THE DEVELOPMENT OF NEW FOOD PRODUCTS

Edited by:

Reema

Delve
PUBLISHING

www.delvepublishing.com

The Development of New Food Products

Reema

Delve Publishing
224 Shoreacres Road
Burlington, ON L7L 2H2
Canada
www.delvepublishing.com
Email: orders@arclereducation.com

ISBN: 978-1-77469-476-3 (Hardcover)

ABOUT THE EDITOR

Dr. (Mrs.) Reema is presently working as Guest Faculty, Department of Home Science, Government Degree College, Katra Chuggupur, Sultanpur, Uttar Pradesh. She graduate from the Deen Dayal Upadhyay Gorakhpur University and completed doctorate from Nehru Gram Bharti Deemed to be University, Prayagraj in the year 2020 with UGC NET, JRF &SRF. She has published four research paper in elite national publications and participated many national & International conferences.

TABLE OF CONTENTS

LIST OF FIGURES

Figure 3.1: In today's volatile markets, food producers must develop new food products in order to stay afloat. Despite the multiple options, the mistake rate for new items is considerable. As a result, a tool for structured food product creation has been created in order to improve the process' success rate.

Figure 3.2: The creation of new products is a lengthy process. When compared to similar items, the product should meet the marketing criteria and have a higher client preference. New goods are classified as "new to the world," "new to the company," "extension to an existing product line," and so on. In today's world, there are several inventions in the food industry. Every day, a slew of new food products enters the market, vying for consumer attention.

Figure 3.3: The initial stage in every product development process is to generate ideas. This necessitates a search for realistic product solutions that can be implemented. It is a critical phase in a company's problem-solving process.

Figure 3.4: Individuals and groups can both benefit from free or lateral thinking. Brainstorming by a group of individuals is frequent in firms that aim to nurture innovation, and it is termed the PIG (Product Idea Generation) group for product ideas to keep it focused on product ideas rather than general brainstorming. Individuals who create their own creative approaches to produce fresh ideas are referred to be inventors. Techniques for improving creativity can be taught to individuals.

Figure 3.5: Idea screening is the next phase in NPD. Idea screening, on the other hand, entails sorting through the ideas to find the best one. We locate several ideas during Idea creation, and we screen them during this step to select the best one. The reason behind this is because product development costs a lot of money in the end, therefore the corporation would rather move through with the ideas that will turn into successful items rather than the ones that will fail.

Figure 3.6: The goal of concept screening is to catch problems early in the development process and avoid them. The earlier faulty ideas are abandoned, the less money is invested and lost.

Figure 3.7: Because every product is not designed for every client or customer group, defining the target market for a product or service must occur early in the product development process. The importance of doing quantitative market research at every stage of the product development process cannot be overstated.

Figure 3.8: Two categories of persons arise when carrying out the innovation process. The first wants all of the paperwork to be completely filled out, as if they were for a school assignment. The second is more concerned with the goals that each step should achieve. Checking every box isn't as important.

Figure 3.9: Concept testing is a great way for a firm to figure out if its product, design, service, or marketing is feasible in the target market and go on with the proper thoughts or ideas. Furthermore, concept testing will identify opportunities and dangers that will benefit from being addressed at the appropriate moment.

Figure 3.10: To stay afloat in this fiercely competitive business, food and beverage companies must step up their game. The marketing plan will emphasize the company's

Figure 6.8: Functional meals will grow more sophisticated and popular, especially when the now-legal cannabinoid CBD is introduced. Larger M&A acquisitions, as well as a slew of smaller ones, will set the course for where Big Food will go next. Packaging and manufacturing will continue to be more environmentally friendly, and businesses will get closer to meeting their sustainability goals.

Figure 6.9: Every new food business hopes to see their product on the shelves of Whole Foods. However, it has caused some well-known firms to go out of business. Contracts will leave you with small profit margins but high order quantities. To expand up, you may need to invest in costly equipment.

Figure 6.10: By 2050, food consumption is anticipated to rise anywhere from 59 percent to 98 percent. This will have a profound impact on agricultural markets that we have never seen before. Farmers throughout the world will need to improve crop output, either by expanding the quantity of agricultural area available for crop production or by improving productivity on existing agricultural lands with fertilizer and irrigation, as well as embracing innovative technologies like precision farming.

Figure 6.11: Climate change is accelerating at an extraordinary speed. A mountain of data implies that global temperatures will remain to rise, and that human activities like fossil fuel burning, deforestation, and agriculture are the primary contributors. These activities emit greenhouse gases (GHGs) such carbon dioxide, methane, and nitrous oxide, which trap the sun's heat and warm the atmosphere, hence the term "greenhouse." Natural processes also emit GHGs; however, the ability of trees, soil, seas, and other sinks (storehouses) to sequester (catch and store) emissions has typically counterbalanced these emissions.

Figure 6.12: In reaction to socioeconomic and demographic shifts, food systems in low- and medium-income nations are undergoing a fundamental transformation. Rising wages, rising urbanization, and the rise of the middle class all result in significant changes in dietary habits and consumer behavior, necessitating governmental and private efforts to promote food market integration. Increased agricultural production has only partially translated into reduced hunger, and it has not been matched by a drop in malnutrition.

Figure 6.13: Food production also has a significant environmental impact. Agricultural production (encompassing pre-production and post-production operations) accounts for 21-37 percent of anthropogenic greenhouse gas emissions; almost 80% of all vulnerable terrestrial bird and animal species are in risk owing to habitat loss due to agricultural growth.

Figure 7.1: A food-based score and incidence of overweight/obesity

Figure 7.2: Systematic review into city interventions to address obesity

Figure 7.3: Consumption of ultra-processed foods and cancer risk

Figure 7.4 Foods That Can Cause Cancer

Figure 7.5: High salt diets contribute to high blood pressure

Figure 7.6: Diet Chart for High Blood Pressure Patient

ABSTRACT

This book seeks to evaluate the opinions of new product development (NPD) researchers and their partners. These incorporate the people who have advanced their theories with specific reference to the food sector and the creation of new food items. This book distinguishes various theories and methods for dealing with new product development. The book offers the structure to understanding and valuing the complexity of food development projects. Additionally, fashions tools for progress in working separately or in groups to make or develop food items and product offerings are discussed in this book. This book likewise offers challenges and prospects for educators, in capstone as well as in different courses. Since product development is by itself an expansive and interdisciplinary undertaking, the author calls upon readers by introducing data and details on the New Product Development process. Readers, students, and researchers will acquire knowledge on the fundamentals of food chemistry, sensory analysis, statistics and food processing through this book. This book will be ideal for food researchers, food engineers, and regulatory professionals working on the development of new food products.

PREFACE

The concepts in this book incorporate and transcend the ideas regularly offered in food science programs. The authors address pragmatic issues, for example, food marketing, product practicality and industry expectations. A large part of the applied technology covered in this is volume comes from consultations with specialists in the sector. While the volume tries to give an audit and outline of data that is useful in the food business, no single publication can cover everything. Subsequently, the book is a stepping-stone and direction for the readers' individual work and research.

The book offers the structure to understanding and valuing the complexity of food development projects. Additionally fashions tools for progress in working separately or in groups to make or develop food items and product offerings are discussed in this book. This book likewise offers challenges and prospects for educators, in capstone as well as in different courses. Since product development is by itself an expansive and interdisciplinary undertaking, the author calls upon readers by introducing data and details on the New Product Development process.

Similarly, the book supports cooperation among food development specialists. The volume assists readers with solidifying what they know in food product development. In each instance the author welcomes readers to apply what they know to building new items and simultaneously gain from others what should be mastered. In this specific case, readers will acquire knowledge on the fundamentals of food chemistry, sensory analysis, statistics and food processing, which are captured in the pages below.

Food product development includes more than simply making the ideal recipe. Organizations should plan widely, work hard, and research extensively to deliver new food products. Before beginning a new development project, it is important to create unique targets and plans that coordinate the future bearing of the business. Organizations participate in new product development with the expectations of gaining new clients, venturing into new geographic business sectors, increasing benefits, raising brand recognition, or expanding portfolios of the overall food industry. Organizations, both big and small, develop a great many new food products every year. The time spent developing new food products varies from a half year to 5 years, contingent upon the level of new innovation and development. For instance, line product development that uses tools already established at an assembling plant typically takes less time to create compared to custom processing lines.

The failure percentage of new products, which is characterized as a product not in the store shelves anymore after five years, can be as much as 90% in certain grocery classes. Bigger organizations depend on product development teams that incorporate food researchers, food engineers, and regulatory professionals, marketing specialists, and buying specialists, whereas smaller organizations may not have a product development division. This book further discusses the steps involved in food product development, and fermented food varieties alongside cases of each, including feasibility obstacles to product commercialization that are the fundamental components to develop effective food products.

CHAPTER 1

FOOD INDUSTRY IN THE WORLD

CONTENTS

1.1 INTRODUCTION

Humans have fought and struggled for existence since the beginning of time when life on Earth started. Throughout human history, humans began cultivating the land, farming, and raising animals to make food more convenient and enjoyable. Later on, individuals began trading and selling their items to swap food for other things or profit from their wares. Food supplies have recently risen dramatically and have continued to evolve. Unlike individuals in the past who had to work for their daily meals, we in the Western world are continuously assaulted with commercials and promotions for food products of all varieties, qualities, and prices. Despite what appears to be a paradox, people in the distant past made quite straightforward judgments about what to eat: they just hunted whichever animal was available at the time to offer the most nutritious meal. Humans consume a diet that is high in lean meats, fruits, and vegetables that they are unaware of. Civilization, industry, technical advancement, automation, economic expansion, and increases in human well-being all played significant parts in the enormous transitions that happened throughout the twentieth century (Yang et al., 2017). As a result, consumers no longer seek out healthier meals, choosing to eat more enticing and appetizing stuff instead. Many of these meals are strong in sugar and carbohydrates but poor in vitamins and minerals. Indeed, the twentieth century saw the largest development in the food industry. Food wholesalers see consistency, concentration, automation, and simplicity as the keys to success. "A&W All American Cuisine," the world's first fast-food chain, was launched in 1922.

In this essay, I'll look at the reasons that have led to this company's long-term success in the cutthroat food market. The food industry was dubious about its future when it was created. Following WWII, the United States and the rest of the globe faced resource and personnel shortages, with only the most well-organized and flexible firms surviving. Companies employed automated and synthetic fertilizers to enhance the output of processed foods, enhancing profits while lowering expenses. This was thought to be the finest way to feed the world's population while simultaneously gratifying customers and producers alike with its low pricing and huge profits at the time. However, this solution did not stay long (Wyrwa & Barska, 2017). When academics and physicians began to realize how people's eating habits and food production systems were hurting their health and the environment, it became evident that something needed to be done to correct the situation. Obesity has grown by 20% in the United States since 1983, with 78.6

million persons classed as overweight or obese. Many younger businesses, such as Panera Bread, have grasped the gravity of the crisis and altered their business models to match shifting customer needs.

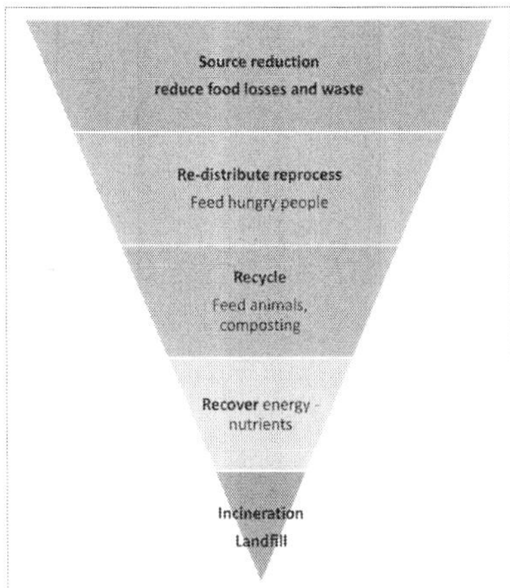

Figure 1.1: Food Security, Safety, and Sustainability.

Source:__https://www.frontiersin.org/files/Articles/487217/fsufs-04-00016-HTML/image_m/fsufs-04-00016-g001.jpg

Over time, the food sector has changed and grown to fulfil the requirements and desires of both customers and consumers. It is distinguished by a complex system of operations that encompasses the production, consumption, and distribution of food commodities worldwide. My dissertation will centre on the evolution of the American food industry from World War II (WWII) to the present. Before World War II, people all across the world relied on locally grown and seasonal foods due to geographical constraints. As a result, further changes happened inside families: as men departed to fight in the war, women were pressed to fulfil their duties as housewives. As their discretionary income grew, so did the amount of time they had to dedicate to housework, increasing in demand for firms that prepared meals ahead of time while they were gone (Wenzl et al., 2006). After the war, processed meals previously utilized by soldiers on the battlefield are now accessible in town and city grocery stores and restaurants. This trend grew even more evident when Richard and Maurice McDonald discovered the value of swiftly manufacturing food, selling it inexpensively, and investing less time and money in obtaining the highest quality food sources and final product.

Figure 1.2: Maurice McDonald.

Source: *https://cdn.britannica.com/35/35835-004-22673F7C/Ray-Kroc.jpg?w=300&h=169&c=crop*

To reduce the cost of the product, the two brothers renovated the food preparation area like a car assembly line. Since McDonald's, there have been other fast food businesses, but none have been as successful as McDonald's. Researchers discovered that the number of illnesses associated with bad eating habits, such as high blood pressure, heart disease, diabetes, arthritis, cancer, and obesity, was on the rise. Several experts, notably Ancel Keys, discovered that saturated fats, as opposed to monounsaturated fats, cause heart issues. In 1968, Senator George McGovern, chairman of the Senate State Committee on Nutrition and Human Needs, initiated an anti-bad-eating campaign. He worked on a range of food assistance programs to develop evidence-based recommendations on the foods that Americans should consume to reduce their risk of heart disease and other chronic illnesses.

When the word "Dietary Goal" was coined in 1977, it elicited significant reactions from the scientific community, the food business, and several public health groups. Soon after, food makers were confronted with a new challenge: as the general public became more aware of the need for appropriate nutrition, the slogan "you are what you eat" began to embody the new attitude toward food in general. People wanted fat-free meals, but the fat was what made the food taste nice in the first place, thus removing it destroyed the flavor (Vidović et al., 2022). As a result, the corporation has had to invest millions of dollars to keep up with client demand while preserving the attraction of its product. Rather than focusing only on the sale of fried products, fast food restaurants began to compete by offering healthier options such as salads, fresh fruit, and organic meals. As a consequence of customers' preference for healthier alternatives, several once-dominant fast food establishments have gone out of business. The first chapter of this dissertation will focus on consumer behavior and new business models that have evolved since World War II.

"Consumer behavior" is the study of how people buy products. Businesses are always looking for new methods to improve their chances of creating and manufacturing the right items for the right consumers at the right time. People establish views before they can even articulate what they are thinking. The process of picking from a list of possibilities grows more meticulous with time, and many people are unaware of it. Other customers, on the other hand, make comparable choices. Companies are founded to produce, deliver, and capture client attention (Vigani et al., 2015). A corporation must have a business model in place, either formally or implicitly, to be successful. This paper reflects management's opinions on what consumers want, how they want it, and the best way to provide those demands—and be paid for and profit from it. The second chapter of my dissertation will look at the growth of the American food business as well as its distinguishing features. Examining the industry's whole history and growth through time will reveal obvious market patterns, changes, and automation.

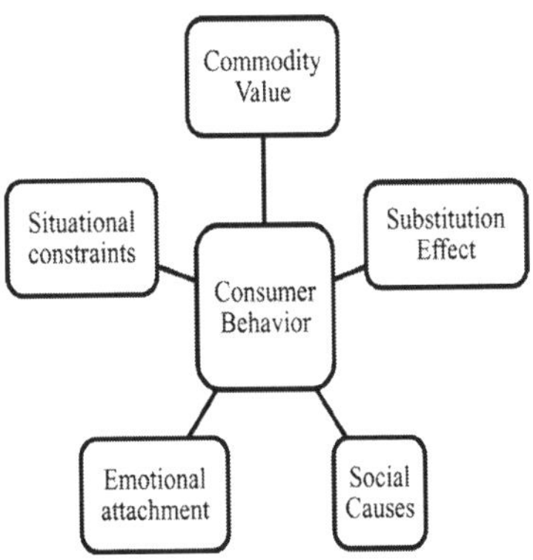

Figure 1.3: The Five Explanatory Variables of Consumer Behavior.

Source:________https://www.researchgate.net/profile/Priyanka-Dey-5/publication/273604946/figure/fig1/AS:391865133223947@1470439336977/The-Five-Explanatory-Variables-of-Consumer-Behaviour_Q320.jpg

Since World War II, we've made considerable modifications to our eating habits. Food advancement required large-scale industrial farming, widespread food preservation through canning or freezing, and a transformation in kitchen technology. After World War II, when women began to replace men who had left the workforce to participate in the war and so had less time to make meals at home, the demand for processed foods increased. Packed food was developed during World War II to satisfy the demands of troops on the front lines, and its broad use in postwar society ushered in a food revolution. The last chapter of my thesis will concentrate on two large food corporations in the United States and throughout the world. Companies such as A&W All American Foods and Panera Bread have built new business models in response to evolving consumer wants and expectations. As seen by the contrast between these two organizations, corporations have acquired various demands throughout time. Following World War II, firms wanted to produce large quantities of food that were quick, economical, flavorful, and appealing (Veeck, 2010). Food is now delivered slowly, healthfully, and organically, thanks to increased health awareness and the importance of nutrition. In my dissertation, I plan to investigate changing consumer behavior and its implications for the food sector. I intend to demonstrate, through a variety of research and data-gathering approaches, how consumer demands for ease of use and enjoyment have increased over time, while their concern for environmental and health-related problems has expanded. Food firms and market leaders have altered dramatically as a result of these innovations.

1.2 CONSUMER BEHAVIOR

Increasing demand for socially and ecologically responsible products reveals that the food industry's trends have evolved dramatically in recent years. When economic expansion started in the early 1950s, people's traditional methods of living were dramatically transformed, which led to an increase in sedentary behavior and an increase in food availability that contributed to a rise in obesity. Obesity has surged as a result, resulting in serious public health concerns. Diet and health have only lately been recognized by governments and the general community. In recent years, customer preferences and expectations have altered. Customers' behavior and purchase decisions are greatly impacted by the link between flavor evaluations and nutritional facts. To complement the male labor recruited during World War II, women began working outside the house (Van Huis, 2020). Housewives who had never worked before the war were able to enhance their families' quality of living. As a result of rising work-related stress and a lack of leisure time, the demand for pre-prepared foods surged considerably. As a consequence of time restrictions, postwar clients ate their meals on the go or during a brief break at work, according to research and analysis. In their place of the traditional family supper, snacks and pre-packaged meals have become the standard. Customers' judgments of pricing, charges, convenience, menu variety, and quality, as well as location, were profoundly impacted by these cultural developments.

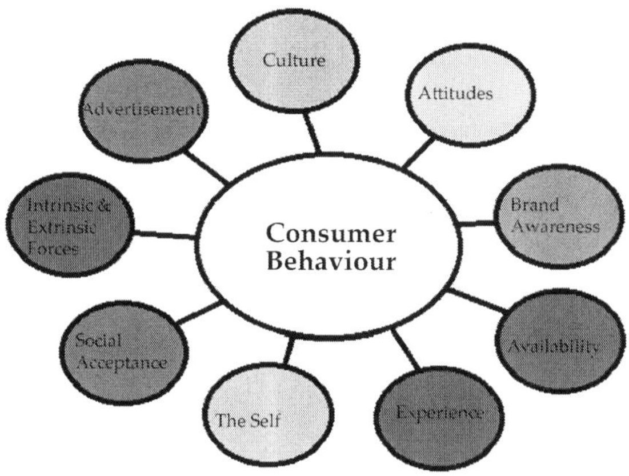

Figure 1.4: Consumer behavior.

Source:_https://upload.wikimedia.org/wikipedia/commons/4/4e/Consumer_behaviour.jpg

Food safety and proper eating habits were unknown to the general public at first, but they have subsequently become more commonly recognized. There was also a lack of healthy behavior among individuals who had been schooled on food and nutrition. It was assumed that to satisfy the newly industrialized economy, Americans needed to make affordable items for all. First and foremost, I aim to research and analyze the shifting marketing and commercial strategies deployed by food firms in light of what I discover about customer behavior. Although consumer behavior analysis was originally brought into marketing research, the predominant focus was on the prospective contribution that psychological studies may make to management choices, patterns have shifted dramatically over time. In the second stage, behavioral studies were described as the theoretical underpinnings for marketing and consumer research, but this endeavor failed (Tarasuk, 2008). A quick survey of recent breakthroughs in behavioral analysis will get things started. For this term, "consumer behavior" refers to the study of how individuals choose (or fail to select) to satisfy their desires via the usage of various sorts of products, services, experiences, or ideas.

Customer behavior helps marketers get a better knowledge of how customers think, feel, and make buying decisions about various things and brands. According to "Consumer Purchasing Behavior," "decision-process" purchases are made. For enterprises to design goods, marketing methods, and business models that can be modified to change client desires and requests, they must do an industrial analysis of the variables driving consumer decision-making. Understanding customer behavior is a tricky subject on the side of economic psychology and marketing science as no two people think or feel the same way (the study of consumer and marketer behavior). The primary challenge is that no one stimulus-response paradigm can account for all human behavior.

When it comes to understanding demand, there have been two major schools of thought: those who focused on building rules to manage the markets and those who focused on evaluating customers' behavior in terms of one purchase. According to economic theory, all customers were rational and aimed to maximize their interests. Individuals who were originally statisticians but moved their attention to psychology to find the rules of consumer choice were also involved. "The Wealth of Nations" by Adam Smith was one of the early studies on consumer behavior, published in 1776. Customer motivation, according to Smith, is founded on the need for self-gratification. It is vital to note that prices fluctuate directly about the amount requested, which is price dependent, and inversely about the quantity supplied, which is also price dependent.

Seventeen years earlier, Smith authored "The Explanation of Moral Sentiments," which gave an opposing theory of consumer behavior to the one previously given in this article. Rather than depending on objective judgments of one's conduct, Smith depended on the assumption that people's acts are directly influenced by their emotions and that the only way to prevent them is to objectively scrutinize one's activities (Sousa et al., 2008). When Edgeworth authored "Mathematical Psychics," he argued that "the essential premise of Economics is the solitary motivator of self-interest for every actor." On the cardinal utility function, Edgeworth wrote that "the bought amounts of each item are arguments, and each good's marginal utility is a diminishing function of the quantity." For most of the 1800s, statisticians depended on primitive methods to identify if two variables were connected and to develop reliable regression models.

Figure 1.5: The Explanation of Moral Sentiments.

Source: https://m.media-amazon.com/images/I/51kUE2mvjTL.jpg

When Marshall produced "Concepts of Economics" in 1890, he made a key addition to the economics canon by presenting the model with the standard "supply and demand" curves that established numerous economic concepts, such as the notion of Marginal Utility. He also developed the phrase "elasticity of demand," which refers to the influence of a 1 per cent change in price on the volume of a product demanded by a customer. Advances in econometrics, as well as demand analysis, were produced simultaneously throughout this period. Traditional knowledge was proven to be fraught with a variety of issues by agricultural economist Moore. Changing the demand and supply curves, as well as the short and long-term equilibrium, was one of the most critical hurdles.

Because of this, statisticians and econometricians came up with new equations that reflect the complete market process better. A sample research idea was also employed in the early 1900s to find inaccurate demand analysis results (Silva et al., 2020). To put it another way, there was no new theory of consumer demand, but rather the consistency and integrity of the demand function and preference mapping were established by a succession of developments and studies. It had been proven by 1939 that the traditional theory of demand analysis had strong underpinnings, as well as its advantages and limitations. Marketing ideas have changed greatly throughout the years. It's a popular fallacy that marketing has its roots in the social sciences, notably economics and anthropology. Behavioral science has had a tremendous impact on marketing, which has kept focused on the individual client over time.

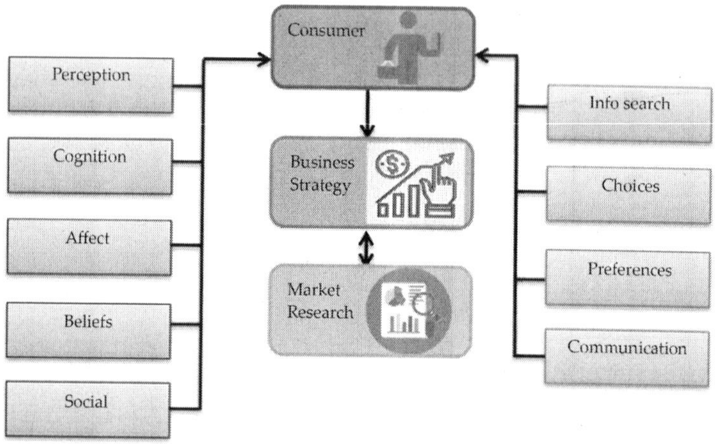

Figure 1.6: Consumer behavior analysis for business development.

Source: https://ars.els-cdn.com/content/image/1-s2.0-S1359178921000458-zr1.jpg

A discipline of behavioral science known as "consumer behavior analysis" focuses on the study and explanation of human consumption. "The consequences it creates, the corresponding incentives and penalties" are utilized to measure performance in this theory. However, it is vital to stress that this new research did not alter long-held assumptions in the area. It was in 1975 that philosopher of science Paul Feyerabend penned "Against Method: Outline of an Anarchist Theory of Knowledge," in which he asserted that there had been an "active interaction between and the great breakthroughs in behavioral psychology that resulted in the 'Consumer Behavior Analysis'." When Berry and Kunkel published "A Behavioral Notion of Retail Image" in the Journal of Marketing in the 1970s, they developed the notion of a behavioral viewpoint on customer purchase choices and marketing operations. Based on sociological research, they investigated behavioral consumer psychology in the article.

To begin with, they needed to build a relationship between marketing and social science for their theoretical study. To sustain and grow demand, marketing is all about "the analysis and enjoyment of consumer-buyers." In 1981, Rothschild and Gaidis published "Behavioral learning theory: its application to marketing and promotions," establishing the significance of the connection between behavioral psychology and marketing (Siegrist, 2008). According to their definitions, human action was a basic consequence that could be changed by both good and bad impacts from other creatures. Radical Behaviorism and Consumer Research Theoretical Promise and Empirical Problems" was published in the International Journal of Research in Marketing in 1988 by Foxall, who explored the major influence of language and verbal conduct on people's behavior and, consequently, decisions. People's everyday lives are continuously touched by a myriad of elements, and consumer psychology gives insight into this.

External (economic, technical, political, cultural, demographic, and natural) and internal (personality, stage of life, attitudes, motives, sentiments, and knowledge) variables impact the consumer's sense of self and behavior daily. Organizational variables (brand name and advertising) also have a substantial influence on the consumer's identity, beliefs, attitudes, and actions. Purchases can be driven by a range of variables, such as a person's own "self-concept," their "nostalgic attachment," their "interdependence," or even their "love" (an emotional connection with the buyer (Sanguansri & Augustin, 2006). The adjectives 'rash' and 'pathological' are used to characterize various forms of consumption. Anti-normative habits include "addictive," "impulsive," "immoral," "illegal," and "compulsive" consuming. Psychology, sociology, anthropology, economics, and marketing management all affect consumer behavior. It's not only about money when it comes to a company's potential to produce money and thrive in today's competitive industry.

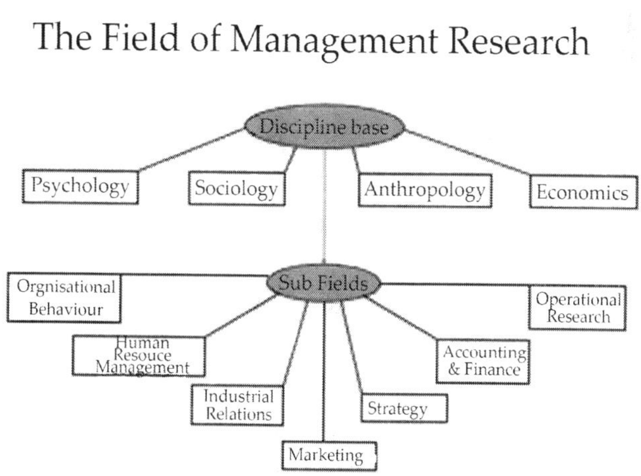

Figure 1.7: Marketing research.

Source: https://image.slidesharecdn.com/chapter3-150108150458-conversion-gate02/85/chapter-3-marketing-research-3-320.jpg?cb=1420751135

Understanding what drives consumer demand, reasons, and preferences is vital to accurately anticipate clients' purchase behaviors. It is vital to note that consumers will commit their limited resources: time, money and effort if a firm is interested in monitoring its customers' purchasing behavior. John Arndt (1976) developed "Consumer Behavior," which he described as "the mental and physical behaviors of an individual or group of persons concerned with final consumption that relate to the purchase and use, and in some cases disposal of economic and consumer items and services." A corporation may invest resources trying to anticipate the number of customers, what they will purchase, when they will buy, and where they will buy.

Assumptions regarding consumer behavior and preferences are generally established by organizations to better understand their consumers' habits and preferences. Numerous models have revealed that firms that understand and satisfy all of the elements of their client market earn more money and make better strategic decisions for consumers; as a result, marketers must emphasize consumer purchasing behavior (Rudolph, 2007). Persons or groups of persons select, acquire and dispose of products and services to suit their aims and preferences. Numerous models of consumer behavior have been established throughout history. Each theory presents a specific insight into customers. The usual models of consumer behavior include the 'Economic Model,' which distinguishes between microeconomic and macroeconomic theories; 'The Learning Model,' 'Psychoanalytical Model,' and 'the sociological Model.' Several recent models exist today, including the 'Nicosia Model' (1966), the 'Howard Sheth Model,' 'Engel Kollat Blackwell Model,' the 'Black Box Model,' and the 'Consumer Decision Process Model.'

1.3 GENERAL CONSUMER BEHAVIOR MODELS

There are differences between traditional and modern approaches to consumer behavior. Four distinct models from the first category are used to understand client behavior:

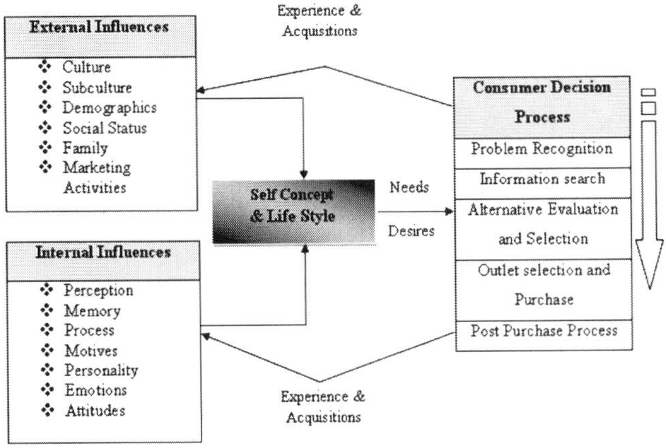

Figure 1.8: The Consumer Behavior Model.

Source: https://www.researchgate.net/profile/Vijayalakshmi-S/publication/322931336/figure/fig1/AS:590613909102597@1517824737883/The-Consumer-Behavior-Model-Hawkins-et-al-1998-External-and-internal-factors.png

Economists were among the first to try to understand consumer behavior. Scientists used to believe that man was completely rational and self-interested. Examining the logical characteristics of human behavior began around 300 years ago. Adam Smith (1723-1790) was an economics pioneer who pioneered the study of consumer behavior. The economic model, which is divided into Micro and Macroeconomic perspectives, describes consumer behavior from an economic standpoint. Both models assume that resources are limited, that requirements are limitless, and that the customer wants to maximize utility at the lowest possible cost. Customers are aware of their options, are capable of evaluating them, and making logical decisions, so the market is thought to have ideal competition. Alfred Marshall developed a utility-based microeconomic model in the early twentieth century, but it fails to explain why and how customers behave in certain ways.

The phrase "Minimum Utility" was coined by the economist (Abbas et al., 2010). Economics is a social science that studies how finite resources are distributed to produce goods that satisfy consumers' insatiable wants and desires. This neoclassical consumer economic theory was based on the assumption that customers would buy a variety of items in varying quantities and at varying prices. It is primarily concerned with the act of purchasing. Marshall's central thesis was that "lowering the price of a product increases the perceived value of the things in the eyes of customers, resulting in more sales"; pricing is the only factor influencing purchase decisions.

Customers are rational and purposeful thinkers who act in their own best interests, according to the model's central assumptions; additional satisfaction from a unit purchased after the initial purchase is less than marginal satisfaction from the initial purchase, and customers have perfect knowledge of the marketplace. Buyers intelligently allocate their limited resources among multiple purchase alternatives to maximize their value and satisfy their desires. The economist, on the other hand, believes that satisfying clients is impossible because their wants and needs are insatiable. The main problem with this model is that economic factors alone cannot explain sales variation, and rationality assumptions are not always feasible. Buyers frequently make passable but not perfect judgments because the time required to complete the best possible study of the various options is not regarded as desirable by the customer. The data presented is frequently skewed.

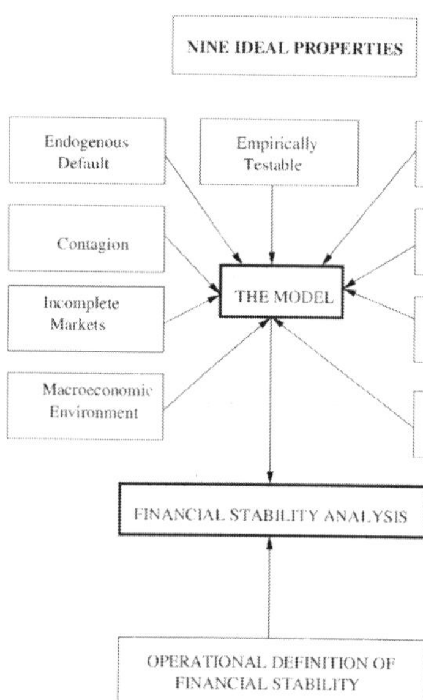

Figure 1.9: Macroeconomic Model.

Source: https://www.researchgate.net/profile/Dairo-Estrada/publication/5143142/figure/fig1/AS:669092390379523@1536535465102/Desirable-characteristics-in-a-macroeconomic-model-for-the-analysis-of-financial-stability.png

The Macroeconomic Model is a monetary model that analyzes the overall economy. In 1949, Harvard University economist James Duesenberry published "Income, Saving, and the Theory of Consumer Behavior," in which he used statistics to show how consumer spending and saving vary across families. In his research, he criticized the preceding strategy for failing to account for psychological effects on client decision-making. It demanded no regard for the idea that others, in particular, could influence people's decisions. Individuals are constantly scrutinizing the purchases of their peers before making their own. Consumption, according to Duesenberry, is determined by the purchasing decisions of members of one's reference group. Peers and social groups generally influence consumption; as people move up (or down) the social ladder, their consumption patterns change (Roy et al., 2009). The hypothesis is known as the relative income hypothesis, and it was developed by Milton Friedman in the twentieth century in his book "A Theory of the Consumption Function." The economist proposed the "permanent income" theory, which holds that consumption is spread out over an agent's lifetime.

According to his theory, a person's purchasing decisions are influenced not only by current income but also by future income expectations, which he refers to as "permanent income," or the average amount they expect to earn over the next several years. Economists and other researchers quickly identified flaws and deficiencies in the economic model. According to experts, numerous behavioral patterns that influenced consumer purchasing decisions but were overlooked by economists could have a psychological basis. Psychologists began to investigate the psyche and consumers' irrational, impulsive, and emotional preferences. The new study looked at how rational people do not always make sound decisions and how environmental factors consistently influence their behavior. This psychological strategy is known as the Pavlovian Learning model because it is based on the research of classical psychologist Ivan Pavlov (1849-1943). Although the formal method of psychological research did not exist until the 1950s and 1960s, studies had been influenced by an economic perspective since the 1940s. Classical Behaviorists are researchers who have taken the second viewpoint and conducted research using a behavioral technique.

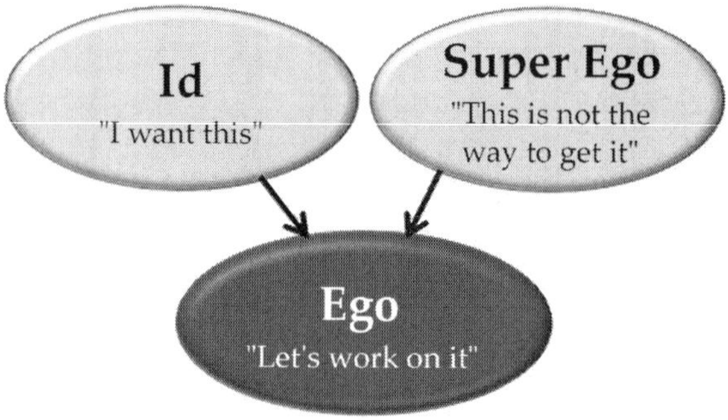

Figure 1.10: Psychoanalytical Model.

Source: https://businessjargons.com/wp-content/uploads/2015/12/psychoanalytic-theory.jpg

Individuals, according to the Learning Model, must repeat the decision-making process regularly. Consumers are problem solvers whose primary goal is to fulfill their preferences and needs. In this paradigm, the client is assumed to adhere to basic learning principles. Learning is defined as a change in behavior that occurs as a result of previous experiences or a comparable event. Learning, according to one of the theory's central concepts, is the result of reasoning, thinking, information processing, and perception. Customer experiences have a significant impact on learning. People never forget their previous experiences or events and continue to learn from them. They gradually learn to connect inputs and reactions, and it becomes second nature to them (Arzani & Ashraf, 2017). Businesses must comprehend how their customers behave in the marketplace. Marketers must understand how customers learn to respond to new marketing scenarios, as well as how they learned and responded to previous analogous occurrences. The learning process is divided into three stages: "Desire," when the individual is confronted with a powerful internal stimulation that, if compelling enough, develops a drive to act to satisfy a need; "Instruction," when the individual is confronted with a powerful internal stimulation that, if compelling enough, develops a drive to act to satisfy a need; and "Instruction," when the individual is confronted The second phase is "Cues," which are external stimuli such as marketing and social media that influence when a client responds. The final stage of this technique, "response," is critical for addressing the individual's compelling need. This series of events will result in the formation of a habit that will characterize one's behavior. Changes in behavior are the result of practice and experience. The model's main flaw appears to be its inadequacy; learning cannot be the primary driver of consumption, and it also fails to account for individual differences and impersonal effects.

The Psychoanalytical Model is concerned with the consumer's personality and reactions. From a psychological standpoint, this theory defines human action. A variety of fundamental causes influence consumer behavior. This concept arose as a result of the work of a group of psychologists who specialized in individual personality. This concept focuses on how purchasers' desires and motivations influence them. Sigmund Freud (1856-1939) pioneered the concept by focusing on people's subconscious and instinctive aspects. People, according to this paradigm, have a complex set of reasons that lead them to make certain purchasing decisions (Balasubramanian & Panigrahi, 2011). Consumers have conscious, utilitarian needs, as well as subconscious reasons that are primarily social and psychological, as well as hedonic in nature, and that people are unable or unwilling to articulate. Many reasons are instinctual (rather than intellectual) or subconscious, and they are symbolic, meaning they are objects or characters that represent an idea or notion. The human personality, according to Freud, has three dimensions: the "ID," which motivates individuals to act; the "Super Ego," which is conscience; and "the Ego," which resolves the conflict between the "ID" and the "Super Ego" and satisfies both. The "ego" is the desire to gratify oneself in a socially acceptable manner.

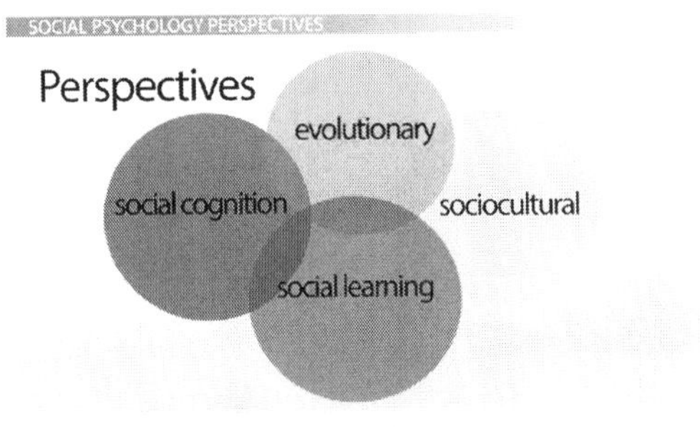

Figure 1.11: Social-psychological paradigm.

Source: https://study.com/cimages/videopreview/videopreview-full/major_perspectives_of_social_psychology_112697.jpg

Individuals, according to the Social-psychological paradigm, are part of an institution called "society," and they are constantly influenced by and influenced by it. People's purchasing habits are changing as a result of the ongoing interaction between civilizations. As a result, utility plays a minor role in purchasing decisions. Social compulsions influence purchase decisions. Individuals are constantly vulnerable to societal pressures emanating from the larger society, subcultures, and other social groups with which they interact. Psychologists have established that certain psychological factors have a significant impact on people's decision-making and must be considered to understand consumer behavior. Perception, learning, and motivation are examples of these aspects (Batista et al., 2013).

1.4 CONTEMPORARY CONSUMER BEHAVIOR MODELS

Consumer behavior was investigated using the Contemporary Models, which were derived from the aforementioned economic, psychological, and sociological areas. Comprehensive models, which take into account a variety of variables, are made up of the following:

The Nicosia model is a simplified, all-variables representation of a more complex model. Francesco M. Nicosia, a consumer behavior expert, devised this strategy. To put it another way, this concept is concerned with how a company and its customers interact. This paradigm is defined as a flow of influences because each component provides input to the next. Customer purchases are frequently influenced by a company's actions, and the firm's target customers are inextricably linked. Customers are influenced to adopt certain attitudes by corporations depending on the terms of a deal. Following that, people will search for or evaluate the product. If this approach is successful, the consumer will be persuaded to buy the product.

The sequence of events is represented by four distinct categories in this model: Field 1 contains information about both the customers and the company (Barreteau et al., 2006). Companies send messages to customers, who then provide information for the next field. A company's message can be conveyed to customers in a variety of ways by using various marketing methods. The first field's early stages are divided into two parts: Sub-Field One includes, among other things, the company's characteristics and customer communication, whereas Sub-Field Two includes the consumer's characteristics, message comprehension, and other inputs. Field 2 is concerned with the investigation and evaluation of various options. This section will entice customers to purchase while also collecting data for the next one. This area is also known as 'Search and Evaluation.' This is the field where the consumer researches the advertised product and the various options; it is also referred to as a "pre-action" field. The client responded in specific ways to the first field's inputs. If the method provides sufficient justification, Field 3 inputs will be generated.

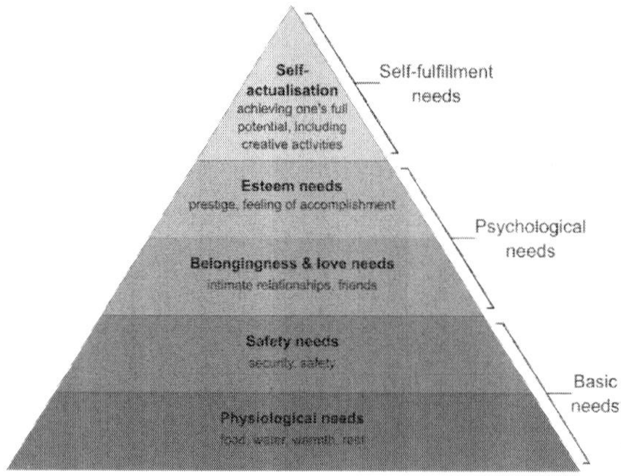

Figure 1.12: Contemporary consumer behavior models.

Source:_https://commercemates.com/wp-content/uploads/2020/09/Consumer-Behaviour-Models-Psychological-Model.png

It is the act of deciding whether or not to purchase something. This is the point at which a person transitions from being inspired to making a purchase. Making a purchase sets in motion a specific pattern of behavior. We're talking about how you used the item in this field, "Consumption." This area's output is information about sales results. The consumer's experience with a particular purchase influences future purchasing decisions... Before purchasing a product, we should consider how we will store and use it. As a result of field 4, the corporation will receive feedback after consumption and will be able to modify and improve its product (Roohinejad et al., 2017). Models like this one have limitations, such as not accounting for the relationship between consumers and businesses and failing to explain internal variables that may influence an individual's personality. This model calls into question some of the model's assumptions and simplifications. The type of client, whether or not there is a relationship between customers and the company, the type of product, and whether or not only the first encounter with a customer is taken into account are all questions that must be answered.

"The Theory of Buyer Behavior" by Howard Sheth, first published in 1969, is one of the most comprehensive theories for understanding a wide range of behaviors. This model is considered to be one of the most comprehensive because it addresses multiple incentives for consumer behavior as well as various stages of the purchase decision. This theory's first integrated model of customer behavior is built on top of the learning model. To explain how people choose products, the model is based on learning theories. With time, the customer's problem-solving activity transforms from a one-on-one endeavor to a habit. Customers make appropriate brand judgments based on the model despite limited knowledge and abilities. This buyer behavior theory is the result of numerous studies in sociology, psychology, and marketing that looked at how people make purchases. The primary goal of this model is to model consumer behavior and its outcomes. The decision-making process is divided into three stages:

Customers who are unfamiliar with a brand's products and have not yet defined their purchasing criteria necessitate in-depth problem-solving (Reis et al., 2014). The customer will conduct extensive research on all of the available brands and their features before making a purchase.

Even though the selection criteria are well-defined, consumers are unsure which brand is the best. Buyers' understanding of the brand and the market remains limited. The factors to consider are obvious, but clients have no idea which solution will be most effective for their specific needs.

The customer's criteria for selection are clear, and he or she has a strong preference for a specific brand. Habitual response behavior: A consumer will purchase something. There are four major components in this comprehensive model: 'Input variables' are: The stimulus comes from the outside world. These elements can convey product or brand characteristics. Significant if they have a direct influence on consumers through brand characteristics (quality, price, service, availability, distinctiveness) or symbolic if they have an indirect influence on consumers and are used by the marketer to represent their product.

The term "output variables" refers to the buyer's potential reactions to stimulus inputs. To make a purchase decision, customers go through a five-step process that includes paying attention, comprehending, forming an opinion, and making a decision. Through various stimuli, an understanding of various brands is developed, which leads to a specific attitude toward the product and a specific desire to purchase it.

Hypotheses are classified into two types: "perceptual" hypotheses and "learning" hypotheses. These are the elements that have an impact on consumer decisions. In the first step, people's reactions and perceptions to the data in the input variable are thoroughly examined (Pereira & Vicente, 2010). It is divided into three parts: a consumer's openness to multiple options: Sensitivity to information

Perceptual Bias: The buyer develops skewed beliefs as a result of his varying perceptions of each brand.

To make a decision, the client requires more information.

The "learning" construct, on the other hand, is based on how the buyer learns and grows as a person and as a consumer. When it comes to brand loyalty, consumers have a variety of perspectives. Six different types of knowledge influence consumer purchasing decisions:

What motivates people to act is referred to as "Motives," and what a buyer believes about a brand's ability to meet their needs is referred to as "Evoked Set."

Consumers, also known as 'Decision Mediators,' evaluate their purchase options using a set of criteria.

'Predisposition' refers to the personal preferences of the Evoked Set.

"Inhibitors" are factors that prevent a customer from purchasing.

Satisfaction is a measure of how well a product's perceived performance meets the customer's expectations.

An exogenous variable is a non-constant external factor that may influence a buyer's decision. External factors include socioeconomic class, culture, organization, and pressure. In this theory, there is no clear distinction between external and internal factors, making it difficult to measure or classify the various forces at work. When it comes to purchasing decisions, the phases of a model do not always follow, nor do they always follow a systematic or sequential approach.

Engel, Kollat, and Blackwell debuted the Engel-Kollat-Blackwell Model in 1968, which Rice renamed the Engel Kollat Miniard Model. It's a consumer behavior model that combines problem solving and education. This graphic depicts these components and subcomponents, as well as the numerous interactions that occur between them. Because both have an impact on consumer decisions and reactions when processed, consumer information elicited in this theory may be related to consumer information elicited in the Howard-Sheth Model (HSM). The model variables depict four categories of consumer decision-making processes (Bellows & Anderson, 2006). The data collection process is divided into two stages: "marketing" (advertising, store displays, etc.) and "non-marketing" (family, friends etc.). "Information input," also known as the "data collection stage," is the first stage.

The next step is "information processing," in which the consumer processes and recalls the information they've already absorbed. As a result, the third step of "Decision processing" includes the following steps: recognizing the need, searching for alternative options, evaluating the options, purchasing, and evaluating the results. The final stage of the decision-making process is influenced by a person's beliefs, attitudes, motivations, and evaluation criteria, as well as his or her lifestyle and information influence. The external variables in this theory are identical to the environmental variables in this model, according to Howard-Sheth's theory. The sheer number of variables and the level of complexity involved is one of the model's major flaws.

The Stimulus-Response Model, or "Black Box," is used to simulate the effects of stimuli on consumer characteristics, decision-making processes, and responses. A customer's attributes and decision-making process, which form a "black box," determine their responses to stimuli. Environmental (PEST variables) factors, which are driven by social factors, are in addition to marketing (product, price, promotion, and location). The black box model, as the name implies, is based on the black box theory of consumer behavior, which emphasizes the relationships between external stimuli and consumer actions rather than the internal processes that occur within the consumer. To identify consumer reactions, marketers must first understand the components of the "Buyers Black Box," which include the buyer's qualities (cultural, social, personal, psychological) and the decision-making process (Blissett & Fogel, 2013). As a result, the consumer's internal processes should be ignored, and the focus should be on the link between stimuli and reactions that are the result of the buyer's conscious and logical choice process.

This does not reflect reality because many decisions are made without conscious thought. Customers begin looking for solutions, such as products and services, once they have identified a problem. Perception is the process by which a person receives, organizes, and interprets information from the world around him or her. "Information processing" refers to an individual's process of receiving, selecting, organizing, and interpreting information to form a meaningful worldview. Because consumers frequently seek information about products and services, researchers can forecast future purchasing decisions.

1.5 CONSUMER DECISION PROCESS

As previously noted, consumer behavior is characterized as "the process consumers go through when making purchases, including the elements that impact their decision and usage"; it covers "the numerous reasons why individuals seek for, acquire, use, and eventually discard things." Consumer behavioral research's principal purpose is to establish how people make purchases. This subject is necessary for businesses to make marketing-related decisions. Engel, Blackwell, and Kollat devised a model of consumer decision-making in 1968. Customers apply five unique stages to make purchasing decisions: 'need awareness,' 'information search,' 'assessment of options,' 'purchasing choice,' and 'post-purchase evaluation.' The purchase represents the culmination of a lengthy decision-making process (Peters et al., 2016). The decision-making process is cyclical, adaptable, and progressive, and it is utilized to make a range of decisions. Numerous elements impact how individuals view the various stages. The type of decision makes a difference in terms of both the number of stages and the length of time necessary to finish them.

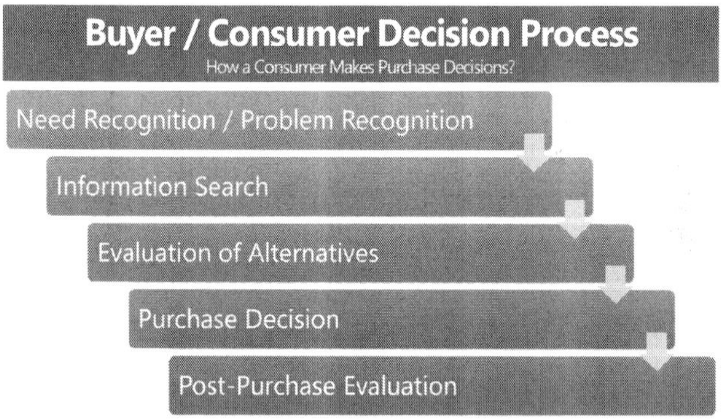

Figure: 1.13: Consumer decision process.

Source: https://i0.wp.com/www.iedunote.com/img/20005/buyer-decision-process-consumer-decision-process.jpg?fit=1920%2C1080&quality=100&ssl=1

The four types of elements that impact purchasing choices are the nature of the goods, the producer's attributes, the buyer's personality, and the atmosphere of the transaction. Businesses attempt to understand these stages to successfully convince clients to purchase certain products or services. Every day, humans must make life-altering choices. We normally make selections without pondering how we got at our opinion regarding which option is the ideal. Certain purchasers acquire identical things, whilst others do not. While humans are not obligated to follow all rules while making basic judgments, certain decisions are so crucial that all processes must be followed. Many persons are uninformed of how long they have been able to get something.

When a customer becomes aware that his degree of enjoyment has gone below an acceptable level, he enters the Consumer Buying Process's "need recognition" stage. A client's perception of a need may be altered by either internal (physiological) or external aspects (e.g. advertising). They must convince someone to initiate the purchase transaction. Maslow grouped desires into five categories: physiological, safety, love and belonging, esteem, and self-actualization (Capitanio et al., 2010). When a person finds which things will meet their expectations, it turns into a 'want.' The second phase is the "information search": once the client has identified his need and desire, he will seek information on the best solution to fill his sense of vacuum. The client can get information either actively or passively. The length of time a customer spends on 'information search' is regulated by the expenditures, the risk involved, and the product's interest. The third step is the "assessment of choices." After acquiring knowledge of different regions, the buyer now has a range of alternatives from which to pick. The individual must study the various possibilities and pick the product that best suits his wants.

To achieve so, he needs to study two parts of the various variables: objective traits such as functionality and subjective ones such as perceived worth. Individuals see qualities differently, which explains why they make distinct conclusions. The fourth phase is the "buying decision." After examining the benefits and downsides of numerous goods and brands, the customer is ready to make his purchase. Unfavorable comments from former clients and an unanticipated external occurrence may both influence the final purchasing choice (change in price, income etc.) The third phase is the "post-purchase evaluation." The apparent performance of the product will be compared to the consumer's expectations (Pereira & Vicente, 2010). Customers who are delighted with a product are more inclined to advocate it to others and to remain loyal to the brand, resulting in higher sales. If the consumer is unsatisfied, he will redo all five phases of the buying process, deleting the previously chosen brand from the list of possibilities. After examining the different consumer behavior models and how they've altered and expanded over time, I'll try to focus on and analyze the firm's business models. Business models have developed greatly throughout time to adapt to an ever-changing market.

1.6 THE BUSINESS MODEL

Businesses must monitor client behavior and the purchase process to produce revenue and profit from sales. To be successful, a firm must have a well-defined plan for capturing value. To develop the value that a firm generates, a complex network of relationships and interactions involving several parties is required. Since 1995, when the notion was initially established, there has been a considerable growth in interest in the subject of business models. Processing, storing, and transmitting information has become significantly less expensive as a result of the introduction of this approach, and it has gained in popularity. Businesses used to follow the same business model; however, choosing a successful market is no longer enough; to compete, you must also build a distinctive business model.

Despite the numerous studies and publications on the issue, no specific definition has been supplied. It's also been called an assertion, a description, an architectural model, a conceptual instrument, and a framework (PF Guiné et al., 2016). However, we may infer that a business model acts as a blueprint for the design and implementation of a company's structure and processes, as well as its physical look. External elements like competition, society, and technical improvements all have an impact on company strategy, which is shown in the shape of a business triangle. You should be aware that there are two main viewpoints on business models: This is how a business runs, and it operates conceptually to see its goals, means, strategy, and techniques for capturing value more simply and transparently than before. A business model goes through five stages of development: The initial phase of published content often focuses on the model's accurate definition and categorization. Following that, the attention switched to determining what made this specific model stand out. The theory's various components were thoroughly outlined during the third phase.

According to Hamel, the "period of revolution" demands fresh and inventive corporate methods. The model must integrate creative and network value. This is a crucial aspect of this unique concept. Afuah and Tucci defined "business model" as "the process through which a firm generates and utilizes its resources to deliver higher value to its consumers while also earning income." In the fourth phase, progress and testing on the model began. The reference model was used in the latter stages of management (O'Shea et al.,

2012). A broad range of business models is used depending on the nature of the product or service being delivered. By identifying the nine basic blocks, the "business model canvas" may be utilized to identify alternative models. There are no distinctions between it and any other model. Businesses may utilize this approach to promote new ideas by pulling inspiration from any of the boxes, and it is inherently related to innovation. Some companies employ business models to foster innovation and skill development. This paradigm may affect the creation of new goods and services in addition to driving creativity. Calia and Maura discussed how innovation may lead to the rethinking of company models. To capitalize on new advancements and technologies, effective business models are required.

To effectively sell their products, businesses must have a solid grasp of customer behavior. Consumer behavior research might include determining which items people want to buy and when they want to buy them. When it comes to evaluating consumer behavior, we examine how consumers decide what they want and why they want it. By getting to know their consumers, businesses may better forecast customer reactions to new goods and services and identify new market possibilities. Customer behavior has a huge influence on company models across a wide variety of sectors. A significant amount of a company's marketing decisions are completely based on customer behavior research (Nicklaus, 2011). Choosing the correct marketing approach may have a significant influence on how buyers consider how to get the most out of a specific product. Marketers must prioritize three separate aspects of customers to effectively sell their products and services: the Personal, the Psychological, and the Social. If a product is geared at teens, such as chips, the company will likely focus on social media; if a product is aimed at homes, such as a vacuum cleaner, the company would likely focus on magazines, television, and posters.

Product development is crucial as a result of customer behavior outcomes. When a corporation detects a market need that has yet to be met, it will introduce a new product or service to fill that need. As a result, to provide a compelling service, firms must acquire as much data as possible from their customers. Service is another key part of a business that must be considered in the previously outlined interaction between consumers and enterprises. Customer service is the consequence of an interaction between a consumer-facing staff and the customer. Customer service has a significant role in the purchasing decisions that clients make. Fast food establishments, for example, are more likely to retain consumers who expect prompt service throughout meals. As a result, to guarantee long-term client loyalty, the firm should continue to train its personnel on how to service consumers fast. Businesses routinely research historical consumer behavior to forecast future sales and avoid product shortages. Customer happiness and loyalty may suffer if a restaurant runs out of the customer's favorite meal, which many consumers do regularly. Consumer behavior has a significant impact on a company's business model; nevertheless, it is critical to investigate how the company's business model affects the company's consumers' behavior.

To thrive, a corporation must understand its consumers and know how to influence their behavior to persuade them to buy its products (Nedovic et al., 2011). Businesses might investigate their customers' wants and goals to persuade them to purchase the items. This can be accomplished in a variety of ways. The first stage for firms is to establish emotional ties with their clients. Eye-catching advertising content is one of the most prevalent, profitable, and quick means of influencing clients. Through television, radio, posters, and other media, businesses' ideas and values are frequently compelling, memorable, and impossible for customers to forget. A consumer is more likely to purchase an item if it is presented in a way that makes him or her feel enthusiastic, amused, or amused. McDonald's has used more than 23 different slogans in its advertisements to promote their goods and has concentrated on the complete experience of the company rather than a single product. "Happy Meals," "I'm having a good time," and phrases like "Get in on the lovin' action ahead of time" are just a few of McDonald's catchphrases.

Businesses have just lately learnt how to influence customer behavior via social responsibility and customer value. A rising number of firms have adopted environmental and charity commitments in recent years, and by informing their customers about their efforts, they have raised demand. Starbucks, one of the most well-known food corporations, is well-known for its environmental care. Starbucks established the "Commitment to Environment" and "Commitment to Communities" programs, which ensured that Starbucks would be environmentally responsible in the communities where it operates and would support local businesses, in addition to the "Commitment to Origins" program, which provided funds and education to small-scale coffee producers, their families, and their communities. As illustrated by the above, the cafeteria's significance in protecting workplace dignity and emphasizing the value of keeping it cannot be emphasized (Monteiro et al., 2010). As a result of current market shifts, the link between customer behavior and company model may alter. Customers' eating habits have shifted considerably, and many companies are seeking to match these new needs by providing more nutritious food. Researchers have shown that firms that fail to keep up with and adapt to altering patterns of customer behavior fall behind and lose their competitive edge.

Corporate customer-focused R&D has resulted in more accurate projections of future trends and larger profit margins than less customer-focused enterprises. Due to the broad diversity of cultural and national food preferences that customers have, restaurants and other food suppliers must understand consumer behavior to modify their approach successfully and efficiently. If a business wants to learn about consumer behavior, it must examine and consider a wide range of elements. Consumer behavior and industrial business models have a significant association, which will be discussed more in the next chapter.

1.7 INDUSTRIALIZATION OF THE FOOD INDUSTRY AND THE BIRTH OF PROCESSED FOODS IN NORTH AMERICA

Industrialization and processing in the food sector led to a significant expansion in the early 1900s. It is necessary to go back in time to understand how much this company has evolved over the years. Modernization, automation, social development, improved quality of life, feminist movements, and World Wars are just a few of the factors that have had an impact on this industry. In the early 1900s, nearly every American's primary occupation was rural farming. As the country became more urban and industrialized, a large number of people from rural areas moved to cities (Morgan & Sonnino, 2010). People had low living standards and little financial means because the economy had yet to reach its full potential. Food production and marketing changed dramatically as a result of the massive political, social, scientific, and economic upheavals that occurred at the turn of the century. Necessities at reasonable prices were required for the new economy to thrive. New transportation techniques had to be developed to keep the food fresh.

Most preservation methods, including drying, salting, and fermenting, were transmitted by ancient civilizations in particular. Canning, for example, was invented by the Dutch navy in the mid-eighteenth century and is still one of the most widely used processes today. The primary goals of food system automation were specialization, standardization, and simplicity. The new concept of specialization was to simplify the manufacturing process so that it could focus on a single product rather than several variants. Market mechanization has resulted in a significant shift in worker occupations. The labor force did not need to learn new skills or acquire new knowledge from employees to perform various production system operations. The primary goal of manufacturing was to produce consistently and efficiently. The economy quickly adopted "the increase in efficiency that results from doing something on a larger scale, such as mass production or bulk purchasing." During these periods of growth, many smaller businesses failed, usually because they couldn't compete with larger, more efficient corporations." One of the most significant achievements of this technology was its ability to produce larger quantities of food with less effort and expense. Customers were drawn to the new market because of its emphasis on cost-cutting (Chaudhry et al., 2008). Soon after, the United States' food supply market was recognized as the most productive and successful in the world. Rapid industrialization harmed many aspects of society, including environmental degradation and animal welfare.

The first half of the twentieth century was characterized by an agrarian society, poverty, and violence. The working class was impoverished, with many families going hungry regularly. Also, infant mortality was quite high. Food was scarce, and the economy experienced a stock market crash that foreshadowed the impending Great Depression during and after World War I. During the 1930s Great Depression, poverty and starvation were widespread. In 1939, while men were drafted into the military, women were forced to fill in for their husbands at work. These changes resulted in some of the world's most significant advancements in economic, technological, and social systems. Following World War II, the American economy could be said to have reached a tipping point. Health, technology, and nutrition all advanced significantly as a result of this terrible historical event. During World War II, food processing began in earnest to make it easier to preserve, store, and transport supplies to the front lines. The military developed microwave ovens and plastic packaging, both of which contributed significantly to the advancement of the food industry. All of these ground-breaking processes are now available to the general public, allowing suppliers to remain profitable while still providing customers with goods that are safe, healthy, diverse, convenient, and appealing.

During this time, food was prepared in such a way that guests could enjoy both safe and elegant items. As a result of the elimination of food restrictions, improvements in health and nutrition education, consumer spending, and automation occurred (Mitterer-Daltoé et al., 2014). What, on the other hand, is food processing? Food commodities are said to go through "any procedure that food commodities go through after they leave the primary producer and before they reach the consumer, who may further prepare or process the food." Processing can occur at various levels of abstraction depending on the sophistication of the technology used. Following World War II, many new food sources became available for use in food processing. There was a shift toward more mechanical, automated, and specialized techniques used by food entrepreneurs. The new industrial leaders' goal was to run viable, profitable businesses by selling highly standardized products at low prices. This market took off in the latter half of the twentieth century. Their new way of life meant they had less time to care for the house and their children, which had a significant impact on their family's finances. Businesses quickly recognized the need for simple, appealing snacks that could be consumed during brief work breaks. Women began to see restaurants as a viable alternative to cooking at home as their standard of living improved. Food preparation was a time-consuming and costly task for women at the time, so feminist groups advocated for food chains and centralized kitchens to alleviate this burden. In the food industry, "fast food" refers to items that can be prepared and delivered quickly. When this type of food was served in restaurants and shops, it was regarded as cheap and of poor quality.

CHAPTER 2

EMERGING TRENDS IN THE FOOD INDUSTRY

CONTENTS

2.1 INTRODUCTION

The food industry is everywhere; it's all the ways we get food into our mouths, from the simplest to the most complicated. This includes everything from growing and harvesting food to preparing and presenting it, as well as the packaging, storage, distribution, and waste management of it. It also includes everything that goes into creating processed foods, including research into nutrition and health.

Figure 2.1: An image showing food production in an industry.

-Source: https://www.google.com/imgres?imgurl=https%3A%2F%2Fwww.foodengineeringmag.com%2Fext%2Fresources%2FIssues%2F2021%2F07-July%2FSOFM-lead-image_900x550.jpg%3F1625575285&imgrefurl=https%3A%2F%2Fwww.foodengineeringmag.com%2Farticles%2F99538-the-state-of-food-manufacturing&tbnid=dr_r4w2cBByV3M&vet=12ahUKEwjR5pSVpvv2AhUJQ_EDHV9PB0gQMygEegUIARCKAQ..i&docid=ia0J9uSc9VyQmM&w=900&h=550&q=food%20industry&ved=2ahUKEwjR5pSVpvv2AhUJQ_EDHV9PB0gQMygEegUIARCKAQ

The food industry is generally divided into three sections: primary processing, secondary processing, and tertiary processing. Primary processing refers to anything involved in farming, fishing, or hunting for food. This is where the raw ingredients come from. Secondary processing takes raw ingredients and turns them into foods that are ready for cooking or further processing. Examples include making eggs into mayonnaise or milk into cheese. Tertiary processing takes those partially processed ingredients and creates finished products like frozen dinners or packaged snacks.

The modern food industry is an evolution of the ancient agricultural system, which dates back thousands of years. As people began to settle down in villages and form civilizations, they started domesticating animals and cultivating crops. These activities formed the basis for today's farming systems. As cities grew in size and importance, so did the food industry. People needed to be able to sell their crops and livestock outside of their small communities, which led to larger agricultural enterprises and extended trading routes. Farmers started doing less of the actual farming themselves, instead hiring laborers to help them produce more crops than ever before. In the early 1900s, food production was largely taken over by large factory farms and processing facilities, which produced significantly larger quantities of food at lower costs. Through economies of scale and other efficiencies, these facilities were able to offer a wider range of foods at lower prices than their smaller predecessors. Some of the world's oldest industries are still a part of today's food industry: baking bread which is thousands of years old, wine making, around 6000 BC, cheese making, around 8000 BC, beer making, around 7000 BC.

2.2 FOOD INDUSTRY TRENDS

The world is becoming more aware of the effects of eating meat and dairy, so innovators are getting creative with plant-based alternatives. The goal is to make the taste and texture of these options as close to real meat as possible, with many companies even going so far as to create "bleeding" burger patties. These vegan options are not only more sustainable; they're also healthier for humans, too. This trend isn't new, but it's growing fast and getting bigger all the time. A good example is Beyond Meat's Beyond Burger; in fact, they're planning on going public in 2019. This isn't just about people who want to avoid meat; it's about what this kind of product can offer in terms of taste and nutrition; and it's also about making these products more affordable and accessible

Figure 2.2: An image showing a brief summary of the food industry trends.

Source: https://www.google.com/imgres?imgurl=https%3A%2F%2Fmms.businesswire.com%2Fmedia%2F20180718005481%2Fen%2F668723%2F5%2FNew_Food_Trends_That_Will_Change_the_Way_the_Food_Industry_Works.jpg&imgrefurl=https%3A%2F%2Fwww.businesswire.com%2Fnews%2Fhome%2F20180718005481%2Fen%2FNew-Food-Trends-That-Will-Change-the-Way-the-Food-Industry-Works-Infiniti-Research&tbnid=Xm5kxBu_WgB6FM&vet=12ahUKEwj6i6PjqPv2AhVagc4BHZS3DvcQMygAegUIARC-AQ..i&docid=YL4FYH9Peo1hHM&w=1250&h=1400&q=food%20industry%20trends&ved=2ahUKEwj6i6PjqPv2AhVagc4BHZS3DvcQMygAegUIARC-AQ

It used to be that if you wanted to start a restaurant, you needed a lot of money and a ton of space. But things are changing; now you can rent a space for your restaurant that's already outfitted with everything you need. This phenomenon is called "plug-and-play," and it allows restaurateurs to focus on the food rather than worrying about how to pay for their appliances or where they'll store them.

Fermentation has been around since the beginning of time. Nowadays it's experiencing a revival with all sorts of new twists; like fermented green strawberries or melon boba tea; that offer all kinds of benefits, like improved digestion and gut health.

Foods that are served at parties and receptions like sliders, french fries, and mini tacos are taking on new heights of fanciness. Instead of serving plain old beef patties, these sliders are being made with ground bison and elk or with pulled pork or even other kinds of meat, such as lamb. The fries have moved from plain to sweet potato and even to non-potato roots like cassava, and the mini tacos can be filled with anything from chicken to shrimp to fish to steak. What's important here is the presentation, which could be in the form of a giant tower or some other visually interesting way, as well as the quality ingredients used.

Every year brings with it some new trends in the food and beverage industry.

2.3 EXPERIENTIAL DINING

Experiential dining is an emerging trend in the food industry that's attracting young people and engaging them. Experiential dining is a restaurant experience that is designed to engage all of your senses and create a lasting impression. They can be in the form of immersive dining, where you eat in an environment that complements the food and helps tell a story; or they can be a multi-sensory experience where you eat dishes that are designed to engage all of your senses. It's no secret that millennials are more interested in consuming experiences than they are things. This generation spends their money on education, travel and other types of experiences rather than purchasing goods. The restaurant industry has been listening to their customers and those customers have spoken. They want experiences when they dine out, which are often shared on social media. Last year, one in five millennials took a food-related vacation, according to recent research from Eventbrite. In addition, 43% of millennials said they're more likely to spend money on a desirable experience or event than material goods. This is why experiential dining is emerging and will continue to grow. The food industry is meeting the demand of millennials' desire for unique dining experiences, with restaurants offering everything from themed meals to multi-course dinners set in unusual locations. In fact, the demand for experiential dining was significant enough for the National Restaurant Association to include it in its What's Hot 2019 Culinary Forecast. The association ranked it as number nine of the top 10 trends for 2019.

Figure 2.3: An illustration of what experiential dining is.

Source: https://www.google.com/imgres?imgurl=https%3A%2F%2Frestaurantindia.s3.ap-south-1.amazonaws.com%2Fs3fs-public%2F2020-02%2FIMG_0437.JPG&imgrefurl=https%3A%2F%2Frestaurant.indianretailer.com%2Farticle%2Fhow-experiential-dining-is-changing-the-face-of-new-age-dining.13547&tbnid=LL9EykE13oL2wM&vet=12ahUKEwiloo2Tqfv2AhWzQ_EDHWIHDR0QMygCegUIARC7AQ..i&docid=L136IbnM5WiI2M&w=1000&h=562&q=Experiential%20Dining%20as%20a%20food%20industry%20trend&ved=2ahUKEwiloo2Tqfv2AhWzQ_EDHWIHDR0QMygCegUIARC7AQ

One of the most exciting things happening in the food industry right now is experiential dining. While high-end restaurants have been using elaborate decorations and artfully-crafted dishes for years, the focus on unique experiences is starting to trickle down to more affordable options. The appeal of experiential dining isn't hard to see. There's nothing like sharing a meal with friends or loved ones and creating memories that will last a lifetime. That's why, if you're looking for a way to make your restaurant stand out, you'll want to consider adding an element of fun or surprise so your customers will remember their experience at your establishment long after they leave.

This can include anything from designing an attractive interior space with lots of natural light and greenery to setting up a small stage where bands play while guests enjoy their meals; there are no limits when it comes to what kinds of things you could do. Another thing that makes these types of restaurants so appealing is the fact that they often offer entertainment as part of their service: DJs spinning tunes on vinyl records during lunch hours or artists painting portraits while people eat dinner.

Whereas as of now a drift within the sense that individuals have embraced nourishment conveyance and takeout, these activities will gotten to be a normal propensity in 2022, permitting benefactors to make their possess encounter at domestic. Eatery chains that had the foreknowledge to offer conveyance alternatives and contactless pick-up earlier to the onset of the widespread have had an upper hand.

2.4 GHOST KITCHENS AND BUFFETS

One of the emerging trends is ghost kitchens taking off. Ghost kitchens are delivery-only restaurants with no dine-in option, and they're also called "cloud kitchens" or "dark kitchens." They don't have to invest in decor or kitchen equipment because they don't need to serve food in person, which means they can cut costs and focus on what really matters: making great food.

Figure 2.4: An image showing an illustration of a ghost kitchen.

Source: https://www.google.com/imgres?imgurl=https%3A%2F%2Fwww.posist.com%2Frestaurant-times%2Fwp-content%2Fuploads%2F2020%2F01%2Fghost-kitchens-1.jpg&imgrefurl=https%3A%2F%2Fwww.posist.com%2Frestaurant-times%2Fcloud-kitchen%2Fdemystifying-cloud-kitchens%2Fstart-a-ghost-kitchen.html&tbnid=cZLV3OB6PQd4MM&vet=12ahUKEwics7POqfv2AhVGiRoKHaUuD8AQMygBegUIARC5AQ..i&docid=g79UbVmImNYXjM&w=741&h=486&q=Ghost%20Kitchens%20and%20Buffets%20as%20a%20food%20industry%20trend&ved=2ahUKEwics7POqfv2AhVGiRoKHaUuD8AQMygBegUIARC5AQ

This type of restaurant is especially popular right now because people still want to eat out, but they want to do it safely. Having staff prepare and deliver the food without having customers come inside helps accomplish that goal. And because you can offer your food at lower prices than a traditional restaurant, ghost kitchens might just be here to stay.

Although restaurants in every niche have been hit hard by COVID-19, the effects of the pandemic have hit some parts of the industry harder than others. One of the most affected areas is buffet dining. Because buffets are known to be risky, even before COVID-19, buffet restaurants have struggled to adapt to a new social distancing world. Many buffet chains have closed permanently or been forced to completely restructure their business model. In contrast, ghost kitchens have seen an uptick in sales during the pandemic. These off-site kitchens, which typically prepare food for delivery only, have been able to meet demand for delivery orders while still keeping workers and customers safe from illness. Buffets are notoriously expensive to operate: you need a lot of space for all the food, every item needs to be kept at precisely the right heat level at all times so that bacteria doesn't grow, and customers generally aren't allowed to touch anything without having their hands sanitized first. The Buffet Industry is struggling, as fewer consumers choose the buffet style of eating. This is partly due to changing hygiene standards, which means buffets aren't being perceived as safe as they once were.

2.5 IMMUNE-BOOSTING FOODS

Immune-boosting foods are trending as more and more people recognize the value of dietary practices that help keep the body in great shape. By now, many of us know that citrus fruits are a great way to give our immune systems a boost. The nutrients in these foods help to fight off illness and improve your overall health, meaning you'll be less likely to catch colds or even suffer from long-term issues like heart disease. This trend is becoming increasingly popular as people learn more about how food affects their bodies.

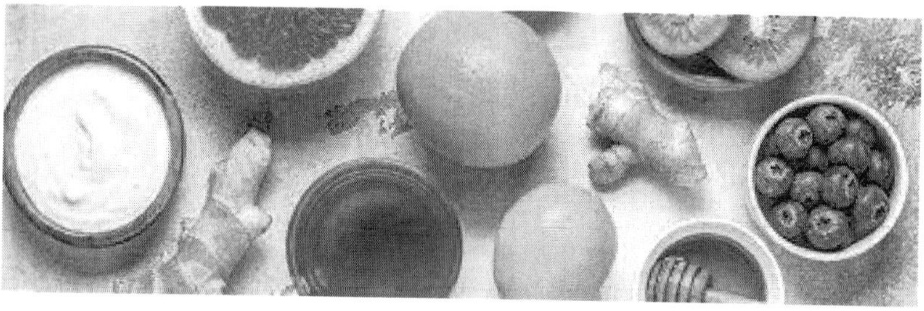

Figure 2.5: An image showing a list of some immune-boosting foods.

Source: https://www.google.com/imgres?imgurl=https%3A%2F%2F2f35da97a9ad36d49db6-4d1987fef3a36cccd5478db4931039f0.ssl.cf3.rackcdn.com%2Fassets%2Fmedia%2F2020%2F06%2FSocialMedia_UK_COVID-19-Im-

mune-health_Blog_1000x305-1.jpg&imgrefurl=https%3A%2F%2Fwww.mintel.com%2Fblog%2Ffood-market-news%2Fwatch-covid-19-boosts-demand-for-immune-health-products&tbnid=6kuUcmMBK0o3cM&vet=12ahUKEwj0o4qCqvv2AhUpi_0HHaz9CkIQMygFegUIARC_AQ..i&docid=52xBjDLyGK2NvM&w=1000&h=305&q=Immune-boosting%20Foods%20as%20a%20food%20industry%20trend&ved=2ahUKEwj0o4qCqvv2AhUpi_0HHaz9CkIQMygFegUIARC_AQ

The food industry is seeing a lot of interesting trends right now, and the rise of immunity-boosting foods is one of the most important. Customers want to know that the food they're eating is doing more than just satisfying their hunger: they want it to make them feel better. And luckily, there are lots of great options out there for people to pick from that can help them do just that. Here are some examples:

Real mushrooms aren't just tasty, they're packed with Vitamin D, which can help boost your immune system. And what's more, you can get all the benefits by adding the mushroom powder to everything from coffee to pasta sauce. The next time you're cooking dinner, add some extra mushroom powder to your pasta sauce and watch as your family's immune systems get stronger with every bite.

Ginger has anti-inflammatory properties as well as other great health benefits, so it's no surprise that people have been putting ginger in everything from tea to baked goods for years now. It also helps boost your immune system. That means not only will you feel less pain when you drink ginger tea or eat gingerbread cookies, but you'll be healthier.

The second is berries. Not only do they taste delicious, but they are also packed with antioxidants that can help protect your body from free radicals; which cause damage at a cellular level. This means they play a key role in preventing disease and increasing longevity. The third is garlic. It has long been known as one of nature's most powerful antibiotics, but now it's also thought to be an effective way to boost immunity. Studies show that people who eat more garlic than average tend to have fewer colds than those who don't consume any at all.

Not shockingly, nourishments that have immune-boosting claims are popping up all over. Indeed items that were pillars for a long time are altering their bundling to call out potential immune-boosting fixings or supplements, such as vitamins A, C, and D, protein, and others. Foods that normally contain these sorts of fixings without adjustment are more engaging to buyers. Whey protein, for case, contains all nine fundamental amino acids to upgrade safe framework work. Leveraging the benefits of whey protein in detail will progress nourishment, flavor, surface, and usefulness.

2.6 MEAL KITS

Meal kits are the latest trend in the food industry, a growing market with a compound annual growth rate of more than 50% between 2017 and 2024. Meal kit services provide their customers with premeasured ingredients and recipes to make it easier for them to cook healthy meals at home. The meal kit industry is gaining popularity as more people become interested in cooking and eating healthy food.

Figure 2.6: An illustration of a meal kit.

Source: https://www.google.com/imgres?imgurl=https%3A%2F%2Fmedia.smallbiztrends.com%2F2019%2F08%2FFitness-food-prep.jpg&imgrefurl=https%3A%2F%2Fsmallbiztrends.com%2F2019%2F12%2Ffood-business-ideas.html&tbnid=Z7oR1QAet5fA5M&vet=12ahUKEwjtq5Wyqvv2AhVKzyoKHbJMDp0QMygeegUIARD4AQ..i&docid=y8aT3dIXvmg1ZM&w=1000&h=560&q=Meal%20Kits%20as%20a%20food%20industry%20trend&ved=2ahUKEwjtq5Wyqvv2AhVKzyoKHbJMDp0QMygeegUIARD4AQ

There are different types of meal kits available today including fresh kits, prepared kits, frozen kits and dehydrated meal kits. Some companies offer a subscription service for these meals while others do not require any subscriptions. Fresh meal kit services like HelloFresh provide all-inclusive packages of ingredients needed to cook a specific dish or meal that are delivered directly to your door without the need for any additional preparation time before cooking them yourself at home. Prepared meals are similar except they come fully cooked so they're ready to eat when they arrive but still require some preparation prior to consumption such as heating up on the stovetop or microwave before being served hot out of the package.

Frozen and dehydrated meal kits differ mainly because they contain less fresh ingredients or have been processed further than traditional grocery store items might have been: they need no refrigeration while also not requiring any additional cooking time.

As popularity continues to rise, meal kits also are becoming a healthier option for consumers. With more consumers being health conscious and seeking out organic produce, consumers are using meal kits for this purpose as well. Additionally, many meal kit delivery services offer environmentally friendly packaging and reusable containers.

2.7 FOOD TRANSPARENCY

Food transparency is changing the way we eat and what we eat. More and more, consumers want to know about the ingredients in their food, where they came from, and how they were made. This is a great thing. Food transparency means that companies are being more open about the origins of their ingredients. We see this most commonly on nutrition labels, where certain items must be specified by law. For example, if a can of beans contains beef broth as an ingredient, it will be stated on the label. Companies may also voluntarily provide information about non-GMO products or organic products.

Figure 2.7: An image showing the logo of an organization supporting food transparency.

Source: https://www.google.com/imgres?imgurl=https%3A%2F%2Ffoodindustryexecutive.com%2Fwp-content%2Fuploads%2F2020%2F11%2FIMI_Infographic_4740x4059px-transparency-triumphs_2020.jpg&imgrefurl=https

%3A%2F%2Ffoodindustryexecutive.com%2F2020%2F11%2Ftransparency-tops-innova-2021-food-trends%2F&tbnid=dxv9KYtcycbJqM&vet=12ahUKEwibxb3hqvv2AhWeE2MBHQKXDA8QMygBegUIARCqAQ..i&docid=_jb8V2WjjSzr7M&w=1517&h=1299&q=Food%20Transparency%20as%20a%20food%20industry%20trend&ved=2ahUKEwibxb3hqvv2AhWeE2MBHQKXDA8QMygBegUIARCqAQ

Transparency can also refer to voluntary labels that explain why a company chooses to use certain ingredients. For example, some companies are labeling their products with no artificial flavors or no artificial colors. It's definitely a consumer-driven change: customers are demanding more information about the food they're eating. But there's also been some pressure from non-governmental groups like the Real Food Campaign. This campaign encourages companies to disclose GMO ingredients on their labels and remove artificial preservatives from their products.

Companies are beginning to take notice of this interest by consumers in the story behind their food, and there's a good reason for that: transparency is what people want. In fact, according to one study, 91% of consumers say it's important for brands to be transparent about their practices. It means that companies have to embrace transparency if they want to stay relevant. It's no longer enough to just make great products, you also have to do things like making sure your business practices are ethical and responsible or ensure safety standards are met. And these days, consumers really care about that stuff.

The idea that consumers have a right to know what's in their food, and that they have a right to know where it comes from. In a recent survey, 67 percent of respondents said they felt it was important to know whether their food came from small-scale farmers or large-scale corporations. And nearly half of those same respondents said that it was important for them to be able to track where their food came from. It's no surprise, then, that more and more startups are popping up with products like "farm-to-table" apps or technologies that let consumers see exactly how much sugar is in their drinks. While these products aren't necessarily changing the way we eat just yet, they do raise some interesting questions about our relationship with food, and what we want from it.

2.8 UPCYCLED AND SUSTAINABLE INGREDIENTS

Access to food is a basic human right, but the food industry is often criticized for its negative impact on the environment. It's estimated that over 30% of the world’s farmland is used to produce food that is never eaten. This waste contributes to deforestation and soil degradation, which in turn increases the risk of flooding, desertification, and climate change. In addition, many farms are located in areas where there is no electricity or running water, so farmers must find ways to use natural resources sustainably; a challenge when it comes to growing crops and raising livestock. To combat these issues and make the food supply chain more sustainable, many companies are turning to upcycled ingredients.

Figure 2.8: An image showing some upcycled ingredients.

Source: https://www.google.com/imgres?imgurl=https%3A%2F%2Fwww.foodbusinessnews.net%2Fext%2Fresources%2F2022%2F02%2F03%2FCaPaoUpcycledCertified_Lead.jpg%3Ft%3D1643912503%26width%3D1080&imgrefurl=https%3A%2F%2Fwww.foodbusinessnews.net%2Farticles%2F20595-trends-aligning-to-accelerate-upcycled-food-and-ingredient-development&tbnid=3c3dNTRB9DzBhM&vet=12ahUKEwi_7dqcq_v2AhXzwgIHHU7PBeQQMygAegUIARCzAQ..i&docid=QWBTVJilZLzvLM&w=1080&h=720&q=Upcycled%20and%20Sustainable%20Ingredients%20as%20a%20food%20industry%20trend&ved=2ahUKEwi_7dqcq_v2AhXzwgIHHU7PBeQQMygAegUIARCzAQ

In light of the increasing awareness of the dangers of climate change, one trend that has emerged in the food industry is a tendency towards upcycling and sourcing sustainably. As consumers become more aware of the environmental impact of the food they eat, companies are beginning to adapt accordingly by using ingredients that have less of an impact on the environment. Upcycling refers to repurposing byproducts that would typically be thrown out into something that has a new use. For example, companies are working to combine waste products from citrus production with their products to create something entirely new. The hope is to reduce waste in food manufacturing as well as help farmers sell more of what they produce.

Sustainable ingredients are those that can be produced without depleting natural resources or damaging our environment. While all upcycled ingredients can be considered sustainable, not all sustainable ingredients are upcycled. Sustainable farming practices such as permaculture and regenerative agriculture fall under the umbrella of sustainable ingredients but aren't necessarily considered upcycled. As more and more companies embrace sustainable and upcycled ingredients, the demand for these products will increase as well. Sourcing sustainably is another way that food companies are looking to reduce their carbon footprint. Sustainably sourced ingredients are often sourced locally or from companies that follow environmentally-friendly practices. Sustainability will be a key differentiator in your market, but it's also an amazing way to get your employees and customers excited about what you're doing.

On the cutting edge of upcycled nourishment initiatives is the generation of whey protein concentrates. Powdered whey protein is an upcycled and economical nourishment fixing looked for after by buyers, particularly those on Keto and other trending diets. At one time, whey was disposed of as portion of the cheese-making prepare, but is presently upcycled into utilitarian and nutritious fixings for soups, sauces, dressings, supper substitution bars, protein supplements, refreshments and more. Not to be overlooked is economical bundling. Nourishment companies are moving from plastic and composites to paper and mono-materials, with numerous brands pointing to deliver 100% recyclable or reusable bundling by 2025.

2.9 REDUCED SODIUM AND SUGAR

The food industry has been reacting to changing tastes over the years, and in the last couple of years, it's been clear that consumers are becoming more health-conscious. That doesn't mean that they want to sacrifice taste or convenience, however; it just means they want a healthier version of what they already like. One way that companies are addressing this is by providing reduced sodium and sugar products. Salt intake is a major factor in cardiovascular health, and high sodium levels may also be associated with an increased risk of stomach cancer; Americans currently get about 3,400 milligrams of sodium per day, on average, the recommended amount is 2,300 milligrams per day for healthy adults.

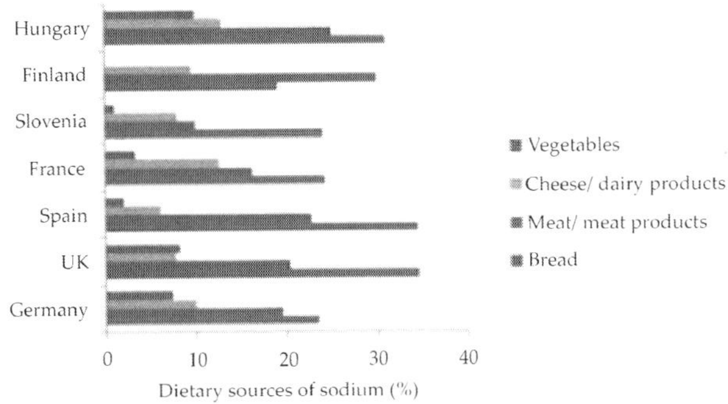

Figure 2.9: An image showing a graph of the rates at which various nations have reduce the intake of sodium and sugar.

Source:https://www.google.com/imgres?imgurl=https%3A%2F%2Fars.els-cdn.com%2Fcontent%2Fimage%2F1-s2.0-S2352364615000024-gr6.jpg&imgrefurl=https%3A%2F%2Fwww.sciencedirect.com%2Fscience%2Farticle%2Fpii%2FS2352364615000024&tbnid=SMqyuV3mFn6w6M&vet=12ahUKEwj4rr3Oq_v2AhVUvSoKHQ83BHgQMygKegUIARDBAQ..i&docid=M1iY_bHakiqJVM&w=491&h=269&q=Reduced%20Sodium%20and%20Sugar%20as%20a%20food%20industry%20trend&ved=2ahUKEwj4rr3Oq_v2AhVUvSoKHQ83BHgQMygKegUIARDBAQ

Excess consumption of sodium and excessive consumption of added sugars both lead to an increased risk for metabolic diseases, such as diabetes, hypertension, heart disease, cancer, and, a variety of other chronic diseases. The good news is that there are a number of ways to lower your intake of both sodium and sugar without sacrificing taste. These include: eating more whole foods like fruits, vegetables, legumes, and nuts; increasing your intake of lean protein; including mono-saturated fats in your diet; and limiting your intake of refined grains, which are high in carbohydrates but low in fiber.

In addition to reducing the amount of sodium in their products, companies are also working to reduce added sugars; which may have an even bigger effect on public health than sodium. The World Health Organization recommends that people get no more than 10% of their calories from added sugars. This can be difficult when you consider how many foods have added sugar—including things like ketchup and white bread.

2.10 CLASSIC FLAVOR COMBINATIONS

Food trends come and go. One moment, it's all about acai bowls and charcoal smoothies; the next, it's hotdogs with macaroni on them. But there's one trend that refuses to budge: classic flavor combinations which work. When you pair foods with complementary flavors, you get something greater than the sum of its parts; and that appeals to the human palate.

Figure 2.10: An image showing the combination of different foods for flavor purposes.

Source: https://www.google.com/imgres?imgurl=https%3A%2F%2Fwww.fona.com%2Fwp-content%2Fuploads%2F2020%2F12%2Ftropical-fruit-salad-papya-mango-pineapple-strawberry-1200x800.jpg&imgrefurl=https%3A%2F%2Fwww.fona.com%2F1220flavorcombos%2F&tbnid=qJzkIWGI4YzNuM&vet=12ahUKEwjOrceWrPv2AhWVVfEDHclWD7IQMygBegQIARB5..i&docid=f8-yIlpccEk2-M&w=1200&h=800&q=Classic%20Flavor%20Combinations%20as%20a%20food%20industry%20trend&ved=2ahUKEwjOrceWrPv2AhWVVfEDHclWD7IQMygBegQIARB5

The food industry has always been known for its ability to innovate. But in the past few years, we've seen a rise in classic flavor combinations being used in new and exciting ways. Whether it's bacon with peanut butter or cinnamon rolls with more cinnamon, chefs and restaurants are striking while the iron is hot, bringing back classic flavors to give customers a sense of familiarity and comfort.

There are two things in flavor profiles: similar ingredients, and contrasting tastes. That can mean a lot of things; you can have sweet flavors in one food (caramel) and a sweet flavor in another food (peanut butter), or you can pair a sweet flavor with a sour counterpart (honey mustard).

Because combining two popular flavors can create a surprising taste that is still recognizable and familiar to consumers. The food industry is not just about creating new things—it's also about re-creating the foods we already know and love. Just like fashion, the food industry has trends that go in and out of style. Some examples of classic flavor combinations include peanut butter and chocolate; Chocolate and caramel; and Apple pie and ice cream. The key is finding contrasts that complement each other.

Each year, there are several radical or uncommon flavor designs that makes the list. Not shockingly, most obscure as quickly as they burst on the scene. Nourishments that incorporate the fifth flavor sensation of umami, in any case, have been on the float for numerous a long time. By and by, creative chefs are trending toward combining those full-bodied, satisfyingly well-off notes of umami with sweet and tart flavor profiles such as citrus. Anticipate to flavor things up in 2022, with hot chile peppers and other abnormal, around the world flavors. Searing sauces, mustards, and other condiments will go with entrees, all with a sign of sweetness from the nectar. Whereas a couple of exceptional flavors may make their check, without address, customers are creatures of reassurance and reach for the commonplace. That was never more veritable than in the midst of the far-reaching when classic reassurance nourishments like soups, savory snacks, and mac and cheese were among the favorites.

2.11 TRADITIONAL PROTEINS

The plant-based nourishment slant isn't essentially going anyplace, but those who were meat-eaters and needed to undertake plant-based proteins will return to their conventional steaks and burgers. Traditionalists and normal shoppers on a budget don't need to give up flavor, surface, and the tactile joys of eating. Perhaps as plant proteins proceed to move forward, they'll try them once more in the future. They'll have a long, tough fight to pick up advertise share against one of the world's driving high-protein fixings; whey protein separate; utilized in protein bars and dietary items. Undeniably, plant-based proteins are developing in ubiquity as an elective to meat, but the most recent investigation appears that whey protein is developing its showcase share in the walk. Whether plant-based bacon will persuade shoppers of its ethics over the genuine thing is however to be seen.

Figure 2.11: An image showing a mixture of traditional proteins.

Source: https://www.google.com/imgres?imgurl=https%3A%2F%2Fwww.greenqueen.com.hk%2Fwp-content%2Fuploads%2F2021%2F10%2FSustainable-Protein-and-Automation-To-Shape-Food-Industry-New-Trend-Report-1.jpg&imgrefurl=https%3A%2F%2Fwww.greenqueen.com.hk%2Fdigitalfoodlab-foodtech-trends-report-2021%2F&tbnid=0jC6PLagnsxBYM&vet=12ahUKEwi5nq7OrPv2AhVM0eAKHWG-Br0QMygNegUIARDRAQ..i&docid=SjR_GXjif7QzeM&w=1200&h=628&q=Traditional%20Proteins%20as%20a%20food%20industry%20trend&ved=2ahUKEwi5nq7OrPv2AhVM0eAKHWG-Br0QMygNegUIARDRAQ

As more and more people start to recognize the value of eating foods that are closer to their natural source, they're looking for ways to incorporate those ingredients into their meals. As a result, they're turning to traditional ingredients; the kinds of foods eaten by our ancestors. They're choosing things like ancient grains and vegetables, which have been eaten for centuries, but have been largely forgotten by the modern diet. They're also reaching for herbs and spices that have long been used in traditional cuisines, but are less popular in modern dishes. We've seen this trend creep into all parts of the food industry: from fast-casual restaurants offering vegan options made with ancient grains to big brand manufacturers offering products made with olive oil instead of corn oil.

When it comes to food and drink, we're all about traditional ingredients. This is something that has become more and more important to us over the last several years, both for our personal health and for the planet too. As much as we love our fancy pre-made meals, there's something so satisfying about knowing exactly where our food comes from; and being able to support the people who grow it.

2.12 CONTINUED SUPPLY CHAIN DISRUPTION

On the heels of major supply chain and diffusing channel unsettling influences that started in spring 2020, several concern remains. In extension to the decreased accessibility of numerous sorts of movement, meat period challenges may proceed to be hampered by the progressing and potential after-effects of COVID-19. Without a doubt in show disdain toward the reality that supplies of these things are returning to commonplace, the openness of talented labor is still misplaced. This might show up as a challenge and by pushed eateries and retailers.

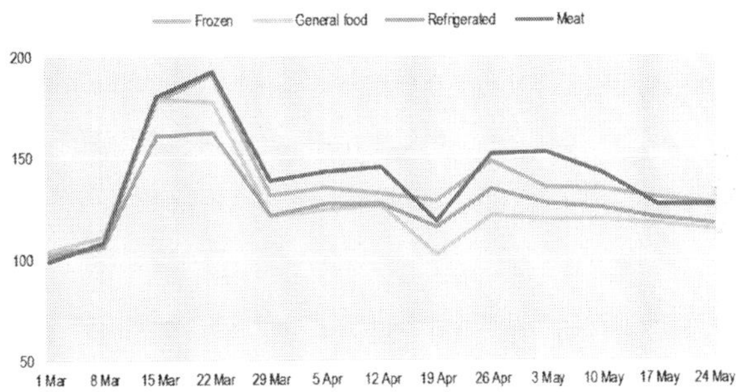

Figure 2.12: An image showing a graph of food supply chain.

Source: https://www.google.com/imgres?imgurl=https%3A%2F%2Fwww.oecd.org%2Fcoronavirus%2Fpolicy-responses%2Ffood-supply-chains-and-covid-19-impacts-and-policy-lessons-71b57aea%2Fimages%2Fimages%2F71b57aea%2Fmedia%2Fimage5.png&imgrefurl=https%3A%2F%2Fwww.oecd.org%2Fcoronavirus%2Fpolicy-responses%2Ffood-supply-chains-and-covid-19-impacts-and-policy-lessons-71b57aea%2F&tbnid=1Mpx7GH9u6l-cM&vet=12ahUKEwjO7-L3rPv2AhWgi_0HHQQ5DRAQMygBegQIARAe..i&docid=S54RqjwLFIdzuM&w=1000&h=504&q=Continued%20Supply%20Chain%20Disruption%20as%20a%20food%20industry%20trend&ved=2ahUKEwjO7-L3rPv2AhWgi_0HHQQ5DRAQMygBegQIARAe

Disruption in the supply chain has been a major trend in the food industry for the past several years, but we may be seeing this trend continue into 2021. Supply chains have been disrupted at every level: from farm to restaurant and everything in between. The pandemic has brought new challenges that food suppliers aren't familiar with, and they're struggling to keep up. In the previous year, many restaurants closed their doors, either temporarily or permanently, due to a lack of revenue. This meant that food suppliers were scrambling to find new customers for their goods, sometimes leaving farmers and ranchers without a place to sell their products at all. This was an issue for both local and national producers. For example, Tyson Foods had to stop production at one of its chicken plants because there were no buyers for the product. Farmers across the country are still struggling to find places to sell their products, which is causing a disruption in the supply chain at every level. It's important that restaurants continue supporting local farmers and ranchers by purchasing their products directly whenever possible instead of through a third-party distributor. The good news is that most food suppliers seem to be optimistic about moving forward. In fact, a recent survey found that nearly 70% of respondents were confident about recovering from supply chain disruptions. That said, the disruptions themselves will continue for some time yet. In the meantime: businesses should think about their customers. If they are selling products directly, consider how you can make your retail location safer for visitors, or if you should shift entirely to online sales. For restaurants, think about if you should keep offering dine-in options, and what restrictions you need to put in place for the safety of both your staff and customers. Considering the diversification of their offerings. We've seen time and time again how difficult it is for businesses to survive on just one source of revenue. Trying something new; though it is not always easy in the best of times.

2.13 PLANT-FORWARD FOCUS

The food industry is always evolving. Each year, new trends and technologies emerge that help culinary professionals adapt to the changing needs of their customers. This year is no different; there are plenty of new trends that chefs and food service managers need to be aware of.

Figure 2.13: An image showing a plant-forward focus menu.

Source: https://www.google.com/imgres?imgurl=https%3A%2F%2Fcdn.winsightmedia.com%2Fplatform%2Ffiles%2Fpublic%2F2019-

12%2Fbackground%2Fplant_forward_RB_1576082495.png%3FVersionId%3DSAwAuJo50yZjOr7yTpx5d5OcG06rfGUS&imgrefurl=https%3A%2F%2Fwww.restaurantbusinessonline.com%2Ffood%2Fhow-plant-forward-trends-provide-menu-innovation-opportunities&tbnid =o2c1XsGur8P5VM&vet =12ahUKEwiy_d6hrfv2AhUL_hQKHQ8YCSMQMygBegUIARCBAQ..i&docid= dt99oGWFNodZfM&w=1800&h=945&q= Plant-Forward % 20Focus%20as%20a%20food%20industry%20trend&ved=2ahUKEwiy_ d6hrfv2AhUL_hQKHQ8YCSMQMygBegUIARCBAQ

One of the most prominent emerging trends this year is an increased focus on p4lant-based foods. Americans are becoming increasingly concerned with the impact they have on the environment, and they're looking for ways to reduce their carbon footprint. This has resulted in an increase in demand for environmentally friendly products such as plant-based foods, which don't require as much resources to produce as meat-based products. Because of this increased interest in plant-forward foods, many restaurants are incorporating more vegetarian options into their menus. Some restaurants are even going all-vegetarian or vegan. There's no sign of this trend slowing down anytime soon: experts predict that by 2025, vegetarianism will be the dominant diet in America.

While veganism is still an important consideration for many diners, chefs are focused on the fact that plant-based diets are better for our environment and overall health than diets heavy on animal products. That's why consumers have taken to calling themselves plant-forward: because they're putting plants at the center of their plates and using meat as more of a condiment than as the star of their meals. We can expect innovative chefs to continue to push this trend into 2019. This trend focuses on incorporating more vegetables and fruits into your daily diet at every meal, which can help reduce the risk for heart disease and stroke while also improving blood pressure levels according to research conducted by Harvard Medical School. With so many delicious recipes out there, it's easier than ever before.

On top of that, there's just a general trend toward cleaner living, whether that means eating less processed foods or trying out veganism for ethical reasons. With all these forces pushing in the same direction, it's not hard to see why so many restaurants are focusing more on plants and less on meat.

2.14 CONSUMING ON-DEMAND

In today's world of instant gratification, consumers are used to getting what they want, when they want it. The emergence of technologies like on-demand music services and streaming video sites has led to a consumer expectation that all content should be available at the touch of a button. This trend has now spread to other industries, with people expecting personalized experiences for all their purchases. One sector where this need for personalization is particularly strong is within the food industry. Consumers have started to demand more than just fast delivery; they expect meals tailored specifically to their preferences.

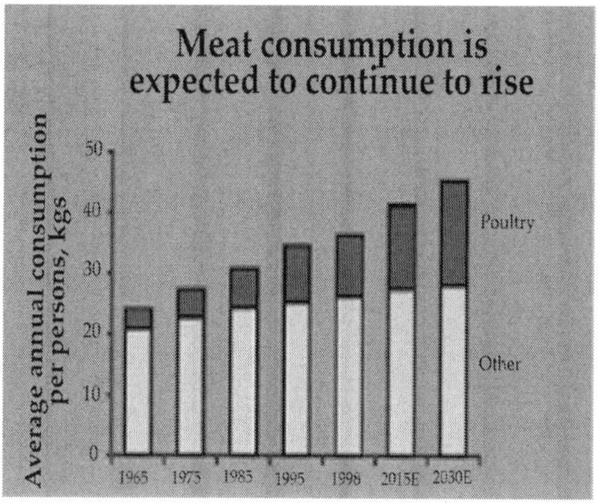

Figure 2.14: A graph showing food demand and consumption rate.

Source: https://www.google.com/imgres?imgurl=http%3A%2F%2Fweb.worldbank.org%2Farchive%2Fwebsite01024%2FWEB%2FIMAGES%2FP52_1.PNG&imgrefurl=http%3A%2F%2Fweb.worldbank.org%2Farchive%2Fwebsite01024%2FWEB%2F0__CO-40.HTM&tbnid=zMdBL-K7CzFlYM&vet=12ahUKEwiv_9DNrfv2AhVUnVwKHZtCAIYQMygGegUIARC-AQ..i&docid=17UyZ0j_ogoOJM&w=364&h=295&q=Consuming%20On-Demand%20as%20a%20food%20industry%20trend&ved=2ahUKEwiv_9DNrfv2AhVUnVwKHZtCAIYQMygGegUIARC-AQ

A number of companies have already started to meet this need. Meal subscription services like Blue Apron allow customers to choose from a list of recipes, which are then delivered directly to their doors along with the fresh ingredients they need to make those meals at home. Other companies focus on creating custom meals based on customer preferences and dietary restrictions, or even blood type. This "consuming on-demand" behavior presents both challenges and opportunities for restaurants looking to stand out in today's market. While some people may choose never to dine out again, others may see it as a way to treat themselves.

This has created an abundance of opportunities for new players in the food industry. For example, The Good Spoon Frozen Yogurt has started offering a monthly subscription service for frozen yogurt. The frozen yogurt is sent to your door every month so that you don't have to go out and purchase it yourself. However, there are also several obstacles in the way of this emerging trend as well. For example, convenience may be one of the biggest reasons for consumers wanting food on-demand, but it comes at a higher price than what would be paid in-store. Also, many consumers prefer shopping in person because they want to see what they're buying with their own eyes before purchasing it. In conclusion, there is always room for innovation in the food industry.

2.15 SCIENCE AND NUTRITION

It has been seen that people are getting more concerned about their health. They have been avoiding food which is not nutritious for them. Therefore, science and nutrition have become an emerging trend in the food industry. People are focusing on their health and also on their future as well. There are some great benefits of eating healthy foods. It is better to eat slowly than to eat fast. Healthy foods should be eaten six times a day for better results. People who eat slowly can live long and happy lives. They are able to control their weight by eating slowly and they also get a good night's rest on a regular basis. There is no denying that fruit is healthy but it is important to make sure that you do not overdo it because too much fruit can cause weight gain.

The food industry is booming, and there's no sign of it slowing down. As we move further into this new decade, we're seeing some trends continue to dominate the space; and others bubble up to take the top spot. For example, sustainability is still a major issue that consumers are concerned about. We've all seen the viral videos of plastic straws in sea turtles' noses, and we're all being encouraged to reduce our carbon footprint. But, as more people focus on their health; particularly as they age, science and nutrition are rising to the top of consumer concerns.

Luckily, food manufacturers are really leaning into this trend, which means there's a whole lot of innovation happening in the field of health and nutrition. Manufacturers are taking advantage of new technologies that allow them to engineer healthier products. They're using new processes to make them more nutritious and organic, and they're marketing those attributes so people can see how the products will benefit their health.

The global market for healthy foods was valued at USD $637 billion in 2018, and it's expected to be worth $704 billion by 2027. And the number of people with diabetes is expected to increase from 415 million in 2015 to 642 million by 2040. As people become more aware of these statistics, they're going to use them to guide their eating habits; and that means food companies without a scientific or nutritional focus are going to see their sales dip unless they adjust their business model.

2.16 MODERN NOSTALGIA

Nostalgia is a big trend in the food industry. It's why you see so many restaurants springing up that look like they were designed in the 1960s, and it's why comfort foods are more popular than ever. As millennials reach adulthood, they begin to crave the tastes of their childhoods, and they're getting the power to recreate those tastes in their own kitchens.

But there's another trend on the rise: modern nostalgia. Modern nostalgia is all about using nostalgia to get back to your roots, while also staying in the present moment. It's the perfect trend for anyone trying to bridge the old with the new, and it's coming at just the right time. In a world where technology is revolutionizing everything we do, it can be hard to feel like you have a firm grip on your identity and heritage. Modern nostalgia helps you reclaim both of those things so you can move forward into the future feeling like yourself. While previous generations might have looked back at fond childhood memories of home-cooked meals and family-style restaurants, millennials are looking back at their college days for the same thing; and many of them don't even have to look back far! The concept of "fast-casual" has emerged in response to this desire for quick, cheap food that tastes good, but doesn't sacrifice quality ingredients or healthfulness for convenience. Popular chains like Chipotle and Panera Bread have led the way here; and now that Millennials are beginning to make up a larger portion of the workforce, they're starting to command more influence over where companies eat together.

2.17 WIDER USE OF AUTOMATION

The food industry is seeing a growing use of automation. The trend began with the largest companies but has now reached small and medium-sized businesses as well. Technologies that were once cost-prohibitive have become affordable, making it possible for more companies to automate their processes and improve their productivity.

Many automated solutions are designed to help food businesses comply with governmental regulations concerning labeling, material tracking, and documentation requirements. Companies that demonstrate an ability to meet these standards are often able to charge higher prices for their products. In addition, they can increase their sales by providing better service and a wider range of options and product availability.

The recent growth in automation within the food industry has also been driven by new technologies that are designed to help companies be more efficient in their operations and reduce waste. Automated solutions include everything from bar code printers and scanners to software systems that can track inventory and monitor quality control throughout a facility. While many of these tools are still relatively new, they are already being used effectively by businesses that want to increase their efficiency and reduce costs. These systems can help a company save time and money in several ways, including reducing errors, increasing production capacity, improving customer service levels, and streamlining operations.

2.18 HIGH-QUALITY FLEXITARIAN FOOD

0The food industry is changing and it's changing fast. Consumers are becoming more interested in high-quality, locally sourced, healthy products. One major trend emerging in the food industry is flexitarianism. Flexitarians are people who follow a plant-based diet but occasionally eat meat. A growing number of people are adopting flexitarianism as a better alternative to vegetarianism and veganism. By reducing meat consumption, flexitarians are helping the environment while improving their health. Along with this trend comes an emerging market for flexitarian products—products that are made to be eaten by both vegetarians and meat-eaters. Companies have begun offering high-quality, delicious products that appeal to flexitarians everywhere. As the food industry continues to grow, we will see more and more companies implementing new strategies to keep up with the ever-changing demands of consumers.

2.19 SUSTAINABLE PRACTICES

Sustainable practices are the new normal for food producers and consumers. The sustainable food industry is currently valued at $1.89 billion, with projections that it will grow to $2.7 billion by 2022. Consumers are increasingly concerned about where their food comes from and how it was produced, which is leading producers to adopt sustainable practices as a matter of survival in the industry. Consumers want to know what they're eating and how they can make sure their food choices are making a positive impact on the world around them. Whether that means small-batch local production or fair-trade coffee, these preferences have become more common than ever before.

Restaurants are doing whatever they can to reduce waste, not just in their kitchens but throughout the entire supply chain. They're looking for ways to limit how much trash they produce, how much energy and water they use, and how much packaging they rely on. They're also thinking about how to preserve natural resources, how stop soil erosion, protect endangered species and habitats, avoid deforestation, and limit carbon emissions. And it's not just about saving the planet; these practices also save money. It turns out that environmental consciousness can be good for your bottom line.

Food producers who can't keep up with consumer expectations are rapidly losing business to companies that have embraced sustainability as a core value. As this trend continues, expect to see even more emphasis on sustainable practices in the food industry in coming years.

2.20 FUNCTIONAL FOOD

Functional foods have been gaining prominence in the food industry for some time now. Functional foods are designed to provide health benefits beyond basic nutrition. They are gaining popularity because they provide health benefits while being convenient and delicious.

Functional foods include probiotic yogurts and kefirs, fermented foods like Kimchi, tasty juices made from fresh fruits and vegetables, and more. These foods may include vitamins, minerals, fiber, protein, herbs, and spices that promote wellness. They range from healthy snacks like granola bars to energy drinks that contain caffeine. Some functional foods are fortified with nutrients like vitamin D or omega 3 fatty acids to promote brain health. The demand for functional foods is expected to keep growing as people look for ways to maintain their health and manage chronic conditions like diabetes.

The emergence of functional foods is a trend in the food industry that is expected to continue in the coming years. Consumers are becoming increasingly concerned with their overall health and wellness, rather than just basic nutrition. Functional foods are one way that they can address these concerns while maintaining the convenience of pre-prepared foods.

2.21 EXPERIMENTAL BREAKFASTS

The world of food is constantly evolving. As tastes change, and we learn more about the impact of our food choices on our bodies and planet, the industry is always working to stay ahead of the curve. One emerging trend in breakfast is experimental flavors.

Figure 2.15: An image showing an experimental breakfast.

Source: https://www.google.com/imgres?imgurl=https%3A%2F%2Fwww.generalmillscf.com%2F-%2Fmedia%2Fimages%2Fnew-industry%2Frestaurant%2Fbusiness-building-ideas%2F6-breakfast-trends-to-wake-up-your-menu-all-day%2Fbreakfast-trends-article-hero.jpg&imgrefurl=https%3A%2F%2Fwww.generalmillscf.com%2Findustries%2Frestaurant%2Fsupport-tool-categories%2Fbusiness-building-ideas%2F6-breakfast-trends-to-wake-up-your-menu-all-day&tbnid=KGeY4ELL5ysUZM&vet=12ahUKEwjthvb3rfv2AhWnS0EAHct3ABAQMygDegUIARC7AQ..i&docid=OJA0DNrJH8zImM&w=656&h=400&q=Experimental%20Breakfasts%20%20as%20a%20food%20industry%20trend&ved=2ahUKEwjthvb3rfv2AhWnS0EAHct3ABAQMygDegUIARC7AQ

More and more restaurants are starting to experiment with their breakfasts. As the number of people who only eat breakfast at restaurants increases, restaurants are seeing a need to innovate with their breakfast menus to keep up with customers' demands. Breakfast is the most important meal of the day, so we shouldn't settle for boring cereal and toast. Restaurants have been working hard to keep up with this demand and make sure that they're offering something tasty and exciting that will get customers out of bed in the morning.

A new trend we've seen is a move away from traditional American breakfast foods like pancakes and waffles towards more international dishes like Indian dosas or Vietnamese pho. This allows them to offer something unique while still offering something that most people are familiar with: eggs. They can also combine different flavors, such as salty bacon with sweet maple syrup or spicy sausage gravy poured over English muffins.

CHAPTER 3

KEY STEPS IN NEW FOOD PRODUCT DEVELOPMENT

CONTENTS

3.1 INTRODUCTION

Developing a food product entail more than just perfecting a recipe. To manufacture new food items, companies must plan carefully, labor diligently, and do substantial research over a long period of time. It is critical to create clear targets and deadlines that incorporate the future direction of the firm before beginning a new development initiative. Companies produce new products with the goal of obtaining new consumers, moving into new geographic areas, improving profitability, raising brand awareness, or increasing market share (Rudolph, 2007). Every year, hundreds upon thousands of new food items are introduced by both major and small businesses. Based on the extent of cutting-edge technologies, the time spent producing new food items might range from 6 months to 5 years.

Figure 3.1: Food Product Development.

Source: https://www.foodbuddies.in/solutions/food-product-development/

For example, developing a line extension that uses existing equipment at a manufacturing plant takes less time than developing a new product that requires a specialized processing line. In some supermarket categories, the failure rate of new items, which is described as a product that is no longer on the shelves after 5 years, can be as significant as 90%. Food engineers, food scientists, regulatory professionals, marketing experts, and buying gurus are all part of a larger company's product development team, whereas smaller enterprises may not even have a development and research team. In order to generate successful goods, smaller businesses may rely extensively on outside resources like independent laboratories and universities (Roy et al., 2009).

The underpinning for both product development processes and product development programs is the product development process. The selection activities in the product development cycle are based on a variety of factors, including business skills and knowledge, ability widely accessible outside the company, risks associated with a lack of experience, the impact of product development on the company's future prospects, and, most importantly, the innovative level.

The ability of a corporation to construct a powerful core and often create fresh recognition is critical to product development success. For product development, there are four main areas where expertise is critical: The cultures of specific geographical locations, their needs and strategies, and a strategy for developing and expanding new products; vital capacity and skills of obtainable raw materials and food handling; high industrial ability and problem-solving skills for establishing new technologies; and What are the numerous systems and management for product advancement? This incorporates a whole technological approach to food product creation, including the association's setting, company resources, expertise, management, procedures, and product development practice.

Idea generation is methodical, informed, and creative. It arises from a thorough understanding of the customer, the market, technology, and the surrounding environment, and it results in novel product, production, and marketing concepts. It generates product concepts in a methodical manner to meet the project's goal and, as a result, the company's business plan (Roohinejad et al., 2017). In the business world, idea generation is planned and not left to chance. Ideas might arise from 'blue sky' research or inventions, but they are methodically turned into innovations in the firm and the marketplace through product development. Idea creation occurs not just in the early phases of creating product concepts, but also throughout the project, including product, package, and process design, as well as marketing strategy development.

Idea creation occurs not just in the early phases of creating product concepts, but also throughout the project, including product, packaging, and process design, as well as marketing strategy development. The scope of idea creation is maintained broad to ensure that no potential breakthroughs are overlooked, but it is narrowed to fit the project's goals. This is a potentially problematic dichotomy.

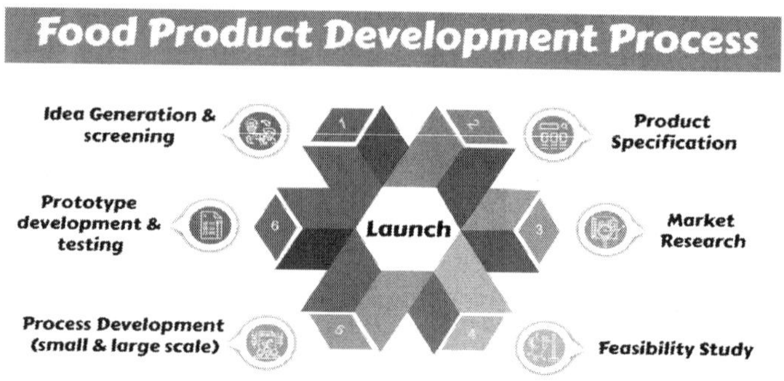

Figure 3.2: The creation of new products is a lengthy process. When compared to similar items, the product should meet the marketing criteria and have a higher client preference. New goods are classified as "new to the world," "new to the company," "extension to an existing product line," and so on. In today's world, there are several inventions in the food industry. Every day, a slew of new food products enters the market, vying for consumer attention.

Source: https://web.facebook.com/FoodResearch/photos /a.116334693344471/322043469440258/?_rdc=1&_rdr

During screening, a large number of concepts are whittled down to a single proposed product, sample product, processing technique, and advertising strategy. The screening begins subjectively and progresses to a quantitative examination of the projected results for the product, manufacturing, market, and finance as more data is received. Throughout the project, there is a continuing cycle of concept generation and screening until the final market launch. A diverse set of concepts is eventually distilled into the ultimate launch strategy (Reis et al., 2014). Controlling these idea creation and screening operations guarantees that no great ideas are lost and that bad ideas are immediately discarded. This is the ideal result, yet it is quite difficult to attain. The scope of concept creation and screening varies depending on the type of innovation and the products; it is minimal for product line extensions, somewhat more for me-too products and product upgrades, and highly wide across the firm for innovation.

Ideas can arise through free brainstorming, rigorously analyzing how the user could use the product, and technological, industrial, and market advances after the project's goal has been set. Using a basic yet rigorous judgment method, these concepts are qualitatively assessed to ensure that they align with the project's goals and restrictions. The chosen ideas are generated into product descriptions and target market descriptions, and the number of ideas is further decreased using a more comprehensive screening process such as inventory screening and economic analysis.

Then there's the consumer-driven creation of product concept ideas, where the focus is on the product advantages; customers and corporate workers steadily limit the quantity of product concepts and produce more precise product concepts. At this level, the review gets more quantitative and precise, and it is based on market research, product costing, and technical evaluations to determine if the product will be a success or a failure in achieving the company's objectives. In product design, commercialization, and launch, the processes of product idea development and screening are continuing; the new product becomes more focused, specific, and quantitative. As a result, everybody engaged in product development has to be able to generate and filter ideas.

3.2 IDEA GENERATION

Companies employ a variety of strategies to come up with fresh product ideas.

Marketing teams may be tasked with the primary formulation of concepts, with researcher input supplemented. Consumer feedback may also provide inspiration. Some businesses may not require this stage, particularly if they thrive on repurposing their competitors' products. A store brand product that is remarkably similar to a named brand product that is sold is an example of this. Corporate idea generation might also include brainstorming sessions with participation from several departments. The key procedures in generating a new food product may be separated into four phases after concept generation: feasibility, screening, test marketing, and commercialization (Pereira & Vicente, 2010).

The interrelationship between the consumer, product and company is the basis of idea generation. These interactions are continually changing, and the environmental condition is undergoing social and technical development as well; comprehending these changes leads to novel goods that meet a demand. From the simplest level, where marketing merely wants a replica of a competitor's new product, to the more complex usage of a new technology, like pressure retention, or a major marketing shift, like the transition from multiple people to single and two-person homes, the product developer must be aware of all of these factors and their interconnections.

The study of interactions is what uncovers and refines product concepts. Is the consumer becoming more conscious of waste packing - can we create an edible or short-term pack?

All new product concepts, whether revolutionary or evolutionary, can only succeed if there is an environment that encourages inventive thinking and the quest for new ideas. Fresh ideas will not be generated if the firm does not foster the process of idea generation (Arzani & Ashraf, 2017). Free idea creation is usually tough for many employees in the firm who have been schooled in logical and methodical thinking. It appears to be a given that a corporation will have very few truly innovative ideas to work on. Product development is frequently required as a result of technology or commercial changes, as well as enhanced understanding.

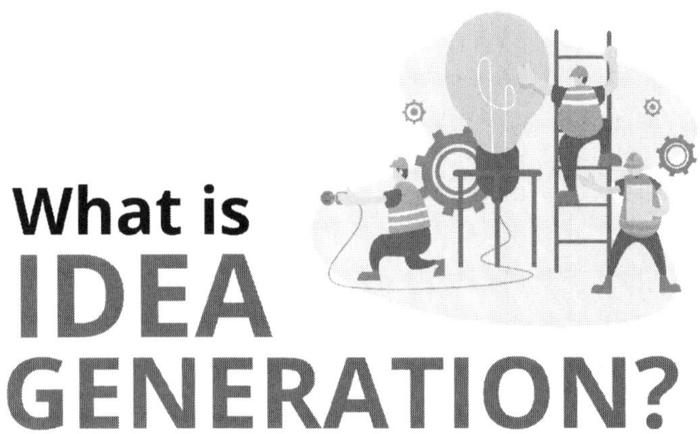

Figure 3.3: The initial stage in every product development process is to generate ideas. This necessitates a search for realistic product solutions that can be implemented. It is a critical phase in a company's problem-solving process.

Source: https://alcorfund.com/insight/idea-generation-2/

As technical and marketing research strive to find ways to improve existing goods or respond to a competitors' products, they are frequently shocked by the utter simplicity of an innovative and successful new product that answers actual customer demands and is growing rapidly in the market. True innovation can provide a new product foundation on which many new evolutionary goods can be built.

Convergent or focused thinking and free or divergent thinking are two strategies for generating ideas, and both are valuable based on the company's product strategy. For the slow development of the product mix, focused, methodical thinking is beneficial. For discontinuous large step-changes, free, lateral thinking is useful. In the food sector, where there is a constant need to produce new goods, concentrated, methodical thinking is emphasized. If food firms want to include inventive new goods in their product mix, they must create an environment that encourages idea development.

There are always challenges in coming up with new ideas, as well as dangers in deciding which path to choose for product development - product enhancement, which seems to be low-risk, requires little study, and is low-cost, or product innovation, which requires considerable research and is high-cost. Making modest product adjustments can lead to slow failure, while picking the revolutionary new product might lead to a quick crash! When deciding on a new product path, expertise and intellect are required (Peters et al., 2016).

Both a 'technological push' and a 'consumer pull' generate ideas. The technological drive stems from an understanding of marketing, processing, and product technology, as well as their scientific foundations. The consumer pull stems from an understanding of the customers and their personal and cultural backgrounds.

The relationship of current goods to user wants, faults in current products, and unmet demands are all part of the consumer needs assessment. With the emergence of 'deprivation foods' low in fat, salt, and sugar, nutritional foods offering supplements of vitamins, minerals, and proteins, functional foods offering physiological advantages and/or reductions in the risk of chronic disease beyond pharmaceutical foods and nutritional needs, providing benefits in health, consumer concerns were a strong pull in the 1990s (Balasubramanian & Panigrahi, 2011).

Technical and scientific literature, corporate R&D scientists, research organizations and universities, firm manufacturing, engineering, and quality assurance employees, and raw material and equipment suppliers are among the technological sources. It usually comes from manufacturing and engineering workers in small businesses, and from the R&D department in larger businesses.

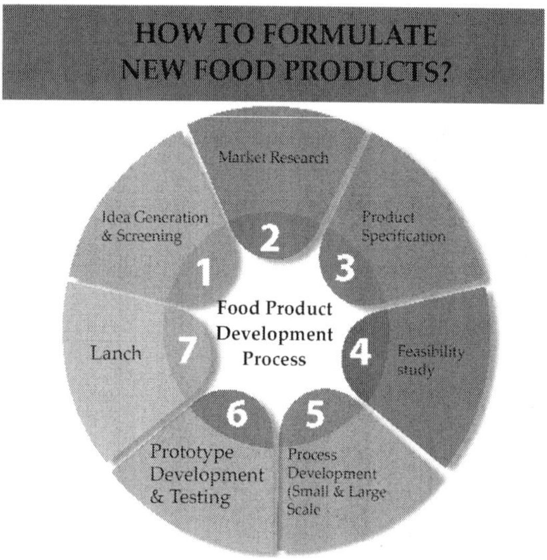

Figure 3.4: Individuals and groups can both benefit from free or lateral thinking. Brainstorming by a group of individuals is frequent in firms that aim to nurture innovation, and it is termed the PIG (Product Idea Generation) group for product ideas to keep it focused on product ideas rather than general brainstorming. Individuals who create their own creative approaches to produce fresh ideas are referred to be inventors. Techniques for improving creativity can be taught to individuals.

Source: https://www.foodresearchlab.com/insights/blog/bakery-confectionery/how-to-formulate-new-food-products/

Competition, international markets, sales journals, customers, market research businesses, advertising agencies, consumer magazines and books distributors (agents, food service, retailers, wholesalers), sales staff, and marketing people in the company are all sources of marketing information. Industry trends, new product debuts, market demands, and market analyses are all included in this data. Retailers may want more brands of a particular product, and they can provide this demand by copying an existing product under their own names. The market may be studied by looking at sales patterns, doing a gap analysis to see if a product is missing, assessing shelf space to see if a line of products needs to be expanded, and comparing the company's product to rival products to see if it needs to be improved.

To find gaps into which the firm may launch a new product, the corporation can create a product matrix comprising their own and rival items, i.e., product categories, product lines, and individual products. The organization conducts market research, such as retail audits and customer surveys, on a regular basis to keep track of the market's 'feel.' Sales patterns and data from grocery sales may now be found and analyzed in great detail. More broad information on economic and business trends may be gained from banks and consultancies, and on societal developments can be found in government or academic social studies publications (Batista et al., 2013).

Food processors, who increasingly provide not just the ingredients, but also the formulation and appropriate market and consumer research. So, where do food processors get their new ingredient ideas? New marine and agricultural goods, as well as studies of the producers' requirements and wishes, as well as fundamental research, are frequently the source of new ideas. Food processors collaborate with marine and agricultural primary producers to develop novel raw materials, and food producers to develop ingredient goods and uses. As a result, they do much of the research in the food sector, typically spending a larger percentage of revenue on R&D than food makers.

Product development, product line expansion, and me-too products are often driven by the market and customers; cost reduction is driven by manufacturing; and innovation is driven by a thorough examination of long-term technical and societal developments. People use many approaches to forecast the result of trends that occur on a regular basis. Historical patterns in population, demographics, household sizes, agricultural productivity, and food production are projected forward in econometric forecasting to anticipate the future (Pereira & Vicente, 2010). Different situations are presented and future results are forecasted in scenario painting; for example, hypothetical scenarios may be: What impact might a reversion to living in small communities in 30 years have on the food industry? or 'China will grow technologically, becoming the world's largest market and exerting significant impact on the worldwide food business' — noodle soup rather than hamburgers as the worldwide food take-away? There are several sources of information for product idea development, and it is critical to recognize them and avoid working with a knowledge base that is too limited.

Individuals and groups can both benefit from free or lateral thinking. Brainstorming by a group of individuals is frequent in firms that aim to nurture innovation, and it's termed the PIG (Product Idea Generation) group for product ideas to keep it focused on product ideas rather than general brainstorming. Individuals who create their own creative approaches to produce fresh ideas are referred to be inventors. Techniques for improving creativity can be taught to individuals.

3.3 IDEA SCREENING

Following the creation of a concept, the processes of product development commence. The most important phase in a design and development endeavor is screening. Thorough product concept testing may help a company decide whether to spend money and time in a business or to abandon it entirely. Project ideas should be in line with the organization's objectives. Throughout the development process, project managers should test concepts to see if the market feasibility of the idea, ingredient accessibility, and regulatory issues have changed. Smaller businesses may use the help of outside firms to do market research.

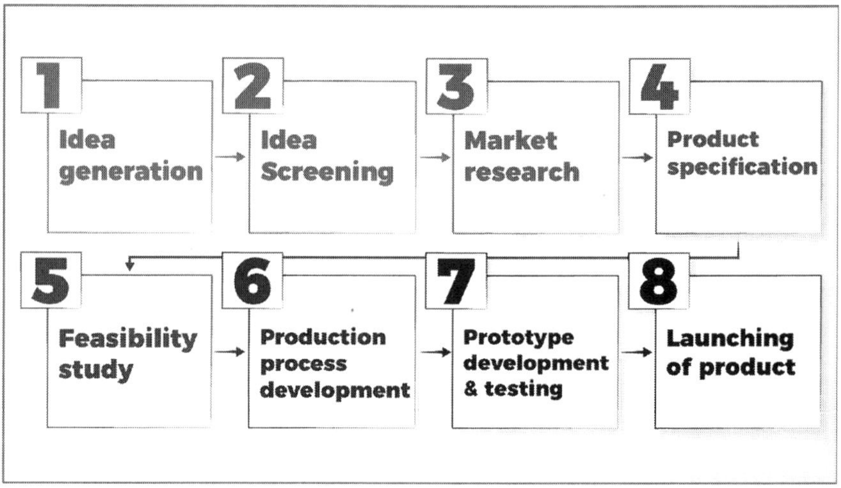

Figure 3.5: Idea screening is the next phase in NPD. Idea screening, on the other hand, entails sorting through the ideas to find the best one. We locate several ideas during Idea creation, and we screen them during this step to select the best one. The reason behind this is because product development costs a lot of money in the end, therefore the corporation would rather move through with the ideas that will turn into successful items rather than the ones that will fail.

Source: https://thefoodtechnologist.com/technology/food-product-development/

The goal of concept screening is to keep the innovative techniques and discard the ones that could fail - much simpler to say than to do in practice! If you're unsure, hold the concept until you get additional information. Idea screening can be focused on the individual's and firm's tacit knowledge, with minimal additional explicit information sought from within or outside the company. However, the goal of subsequent examinations is to gather the information needed to make quantifiable, objective conclusions (PF Guiné et al., 2016).

Product idea summaries or ideas, screening parameters, and screening methodologies are all included in idea screening. Product concept descriptions must be written in a way that everyone participating in the screening process understands and evaluates in the same way. The selection of screening parameters is, of course, critical; the criteria clearly impact the direction of choosing. Finally, the individuals who do the screening, as well as the procedures they employ, have an impact on the screening outcomes. These should be concise and straightforward. A clear explanation of the product, its application, the target market group, the connection to the company's current goods, and the connection to competitor products are among them.

A line of liqueur ice creams, for example, may be a new product concept for an ice cream firm. Is it a liqueur flavor or does it have liqueur droplets imbedded in it when you say liqueur ice cream? Is it intended for affluent diners at leisure or in restaurants? Is it supposed to be the pinnacle of the company's ice cream line? Are specialty ice creams prepared at high-end restaurants the closest competitors? Should members of the community, such as nondrinkers or religious organizations, expect problems, or will there be legalities with excise duties?

The initial description is normally kept wide so that ideas about the product, market, and technology may be continuously researched, but there is a requirement to focus the concept in a certain direction so that the individuals involved do not go in too many different areas at once. The emphasis narrows as the product idea progresses from a product concept description to a concise product idea concept, to the finished product idea, to the product design requirements, and finally to the product specifications (O'Shea et al., 2012). 'A line of simple based ice-creams with small jellies comprising concentrated liqueurs, targeted as a present to be given to dinner parties, offered through higher-class supermarkets,' according to one product concept description for the liqueur ice-cream.

The project goal and project restrictions are the strong screening elements with which the products concept must accord. The company's ultimate goals usually take priority over other considerations. A product concept, no matter how great in isolation, will be dismissed if it does not align with the company's overall business plan, particularly the product strategy. There may be a great product concept that will revolutionize the company's business approach, but it must be extracted from the project concepts and sent back to the top management department. This product proposal must be evaluated in terms of its size and fit for the organization, and senior management choices must be made inside the company.

The project's restrictions, which were specified at the start, are also crucial screening considerations. For a variety of reasons, a product may be dropped: it does not comply with food rules; there is insufficient funding to develop or produce it; or the managing director dislikes it. Although the parameters employed in screening must be as impartial as feasible, subjective judgements are occasionally made.

There are other significant reasons, and these are only a handful that frequently occur. The type of ideas, the firm and its assets, the company's environment, and the level of creativity all influence the selection of screening variables. There may be numerous criteria, but it is impossible to utilize all of them, thus the elements are prioritized in significance and only the most important are picked in the initial screening, with others being examined afterwards. There are four levels of importance for factors: vital, most important, important, and minor (Nicklaus, 2011).

Typically, screening is done in phases. In the early phases, simple approaches may be utilized, with more extensive screening taking place when the number of goods is decreased and more technical, marketing, and financial data becomes available. The quantity of product ideas decreases as screening progresses, but the quantity of knowledge grows, the variety of screening factors rises, and the accuracy increases.

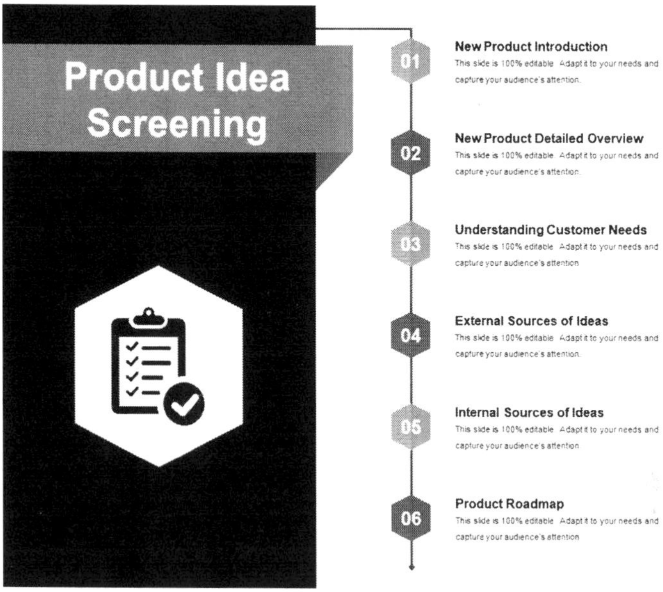

Figure 3.6: The goal of concept screening is to catch problems early in the development process and avoid them. The earlier faulty ideas are abandoned, the less money is invested and lost.

Source: https://www.slideteam.net/product-idea-screening-presentation-images.html

First, the ideas are evaluated to determine if they are consistent with the goal, restrictions, and any other critical variables in a progressive, pass/fail screening, whereby the product concept is evaluated against each critical component in turn, with a simple pass/fail judgment. If they completely fail, they will be removed from consideration. In a checklist screen, the remaining ideas are evaluated against each other on the crucial factors, the scores are totaled to create a total score, and the lowest ranked ideas are eliminated (Nedovic et al., 2011).

Depending on the precision of the information, scales of 0 to 5 or 0 to 10 can be employed. Instead of offering a single rating on a component for the product concept, the probabilities of reaching the various values on the scale are projected using probability screening. At this juncture, extensive data on prices, manufacturing or supply costs, market potentials, and investment costs is gathered, and an economic analysis is performed to determine which product has the highest predicted total sales, financial gains, return on capital, or another financial criterion. This economic analysis is repeated at various phases of the project, along with a review of all the criteria that were considered in previous screens. At the conclusion of the product concept screening, minor variables are added on a checklist to ensure that they are addressed in the finalized product concept. These criteria, and also the items, may be used to screen packaging and advertising.

A wide range of persons are involved in product screening. The preliminary screening is done by firm employees using a basic pass/fail system based on the goal, restrictions, and other critical considerations. This can be done by a single individual, although it is ideal to have representation from many departments (Monteiro et al., 2010). Following the creation of product descriptions, it is critical to include the consumer, or in the case of industrial and restaurant work marketing, the customer. Initially, focus groups are used, but as the number of items decreases, market or consumer surveys on three to five goods are conducted to provide quantitative data on acceptability and commercial potential.

Employees from the company are also participate, either in groups or through a survey. Individual personnel from various parts of the firm may frequently be valuable in scoring product ideas on a variety of aspects and then bringing them together. This draws attention to any potential issues that may arise in the future. You can utilize a checklist or probability approaches. At the end of the first stage, senior management conducts a final screening depending on feasibility studies of the finished product idea.

Throughout the project, concept creation and screening are employed, but the initial product idea generation and screening is the most crucial and has been highlighted in this chapter. All concept generation and screening should be innovative, informed, and methodical. This is a difficult mix to optimize, but it is critical since it leads to the launch of creative and successful items on the market (Morgan & Sonnino, 2010). In the firm, creative individuals are recognized, and an environment is created to support their creativity, resulting in the development of a solid knowledge base in marketing, technology, and customers. Customers for meals and ingredients are also engaged in concept creation and screening to ensure that their requirements and desires are met. Various operating sections of the organization, as well as product design and R&D professionals, are all actively participating. The procedure is organized in such a way that all of these persons are involved, and the outputs are developed and optimized.

3.4 CONCEPT DEVELOPMENT AND TESTING

3.4.1 Concept Development

Regulations, technology, and economics are all factors to consider when developing content for a business. The tools will be accessible to address first issues of achievability that may be raised at any moment throughout the development process by forming an interdepartmental team.

Firms must be aware of the state and/or federal agencies that govern a product before beginning a project. State authorities control items supplied locally in general. Depending on the kind of food, a product crossing state boundaries is within the authority of the US Food and Drug Administration (FDA) or the US Department of Agriculture (USDA). Small food processing enterprises are permitted in certain states, but the processing section must be isolated from the living spaces by strong walls, and there must be no direct access from the living area to the food manufacturing area (Mitterer-Daltoé et al., 2014). Because some local governments restrict in-home commercial food production, it's critical to be informed of local zoning rules.

A processed, cooked, or raw, edible material, ice, drink, or component used or intended for use or sale in whole or in portion for human ingestion, including chewing gum, is referred to as food. Individual food product requirements are established in the code of federal rules, which are referred to as standards of identification. In order to utilize a certain name, a product must satisfy certain requirements. To be labeled as "applesauce," for instance, certain criteria on ingredient inclusion must be followed (21 CFR 145.110). If unsweetened, "applesauce" must have a soluble solids concentration of at least 9%, and 16.5 percent if sweeteners are added (as determined by a refractometer).

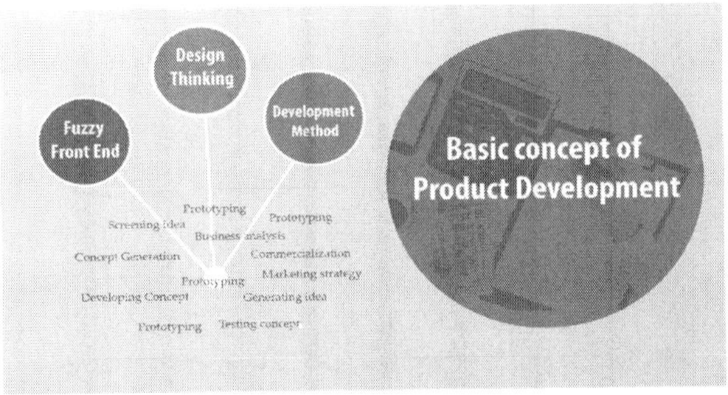

Figure 3.7: Because every product is not designed for every client or customer group, defining the target market for a product or service must occur early in the

product development process. The importance of doing quantitative market research at every stage of the product development process cannot be overstated.

Source: https://www.basic-concept.com/c/basic-concept-of-product-development

The FDA and the USDA each create papers named "Food Labeling Guide" and "The Food Policy and Labeling Guide," respectively, that lay out the guiding principles for labeling items that fall under their control. The USDA Food Safety Inspection Service regulates poultry and meat products that contain more than 3 percent fresh meat or at least 2 percent cooked poultry and are intended for interstate commerce (FSIS). Except for seafood, which is governed by both the Department of Commerce and the FDA, the FDA controls all other food items (Matser et al., 2004).

Jams, jellies, and barbeque sauces, for example, frequently have pH levels that are low enough to avoid these rules. Low-acid food laws apply to most pickled foods and canned vegetables. Other regulated areas that need to be addressed come into two categories: health and economic protections. Food additives, natural toxicants, adulteration, residues, and unclean processing or storage procedures are all protected by health regulations. Labeling difficulties, particularly with regard to misleading or deceptive assertions, and net contents are examples of economic protections.

The required equipment, facilities, and methods to create a product must be developed before a new food product may be launched. When products are discovered to be technologically impractical, the project should be abandoned.

In order to determine the optimal combination of components, processing parameters, and packaging choices to make the desired product, a variety of ingredients, processing settings, and packaging options will be used. This phase necessitates thorough statistical analysis and meticulous record-keeping. Following some preliminary testing, an experimental design will reduce the number of prototypes required, saving time and money (Barreteau et al., 2006). A laboratory notebook should be used to record all calculations and experiments. Each submission should include all pertinent information. This is useful when projects are momentarily stalled, last for an extended period of time, or are handed on to other developers inside the company.

When selecting ingredients, think about whether the product will be accessible all year or only during certain seasons. To verify quality and cost efficiency, product developers may often explore many suppliers of the same product. To meet their demands, larger organizations may need to source many suppliers of the same product. If more than one source is utilized, all vendors must adhere to strict product criteria.

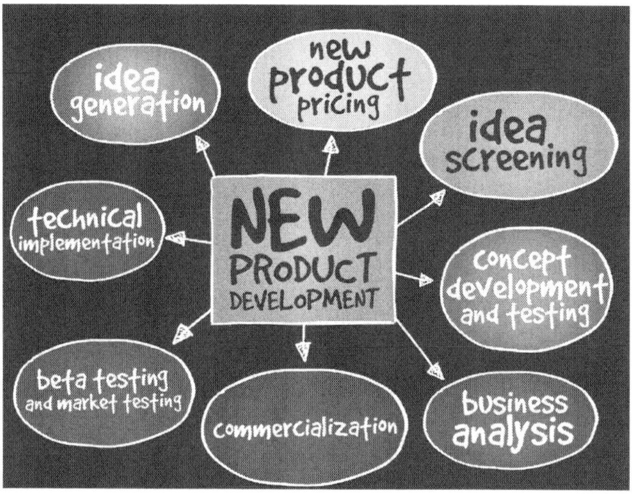

Figure 3.8: Two categories of persons arise when carrying out the innovation process. The first wants all of the paperwork to be completely filled out, as if they were for a school assignment. The second is more concerned with the goals that each step should achieve. Checking every box isn't as important.

Source: https://theaiminstitute.com/product-development/new-product-development-process/

What technology do you already have if you have an established facility? If at all feasible, companies aim to manufacture freshly created items on previously purchased equipment. Because purchasing new equipment is an expensive investment, many product innovation projects focus on extending product lines utilizing existing equipment and facilities (Martelli et al., 2020). As a result, product designers should be aware of the equipment accessible at the site where the final product will be manufactured. If the product will be manufactured in many plants, the focus should be on the disparities in equipment available and how they may be reconciled to create economically comparable goods.

It is important to think about the processing facility that is available. The sewage systems and water supply will need to be evaluated if the firm has purchased a new facility. It is important to consider the conditions inside the plant, such as relative humidity and temperature regulation.

Facilities in places with heat and high humidity in the summer that do not have these circumstances under control may have to change their working parameters in order to produce quality meals. Packaging plays a vital role in a consumer's decision to buy a product, especially for first-time buyers. It's crucial to think about how customers will perceive the packaging and if it will communicate the product's quality goals, such as whether it's a high-end generic grade and premium product. Package engineers, product developers and marketing, should think about the packaging materials used on rivals' products and how they may differentiate themselves.

Frozen and chilled products, for example, have unique distribution requirements. The expense of special events should be considered by businesses. The area in which the item will be offered is another factor to consider. Is the product going to be distributed nationally or regionally? The packing requirements may be influenced by the distribution radius (Lusk & Briggeman, 2009).

The shelf life of a product is determined by how long it retains its quality as viewed by customers. When considering distribution routes, a product's shelf life is critical. Accelerated or real-time testing can be used to evaluate shelf life.

New product creators should think about their goods' safety hazards. The history of outbreaks and reported safety hazards associated with specific product categories might provide insight into risk factors that processors should be aware of. Peanut butter manufacturers, for example, must apply controls to test for probable salmonella contamination following a big epidemic in early 2009 that sickened over 400 people and resulted in at least five fatalities. Some items are vulnerable to spoiling and harmful microbe development. Consumers are also at risk from allergens and physical pollutants such as metal shavings from manufacturing equipment.

Before a food product can be manufactured for sale, it is necessary to have a thorough awareness of all manufacturing and marketing expenses. Prior to manufacturing, a thorough cost study should be performed. Fixed and variable costs are the two categories of expenses to consider. Annual fixed expenses are those that do not fluctuate from year to year, regardless of output levels. These expenditures include things like equipment, buildings, property taxes, and other things that are not affected by production changes (Lamine et al., 2012). Hired labor, raw materials, packaging materials, gasoline, energy, utilities, and other commodities utilized during production are examples of variable costs. Prior to test commercialization and marketing, variable costs should be thoroughly analyzed in order to establish a unit pricing that allows the new product to earn a profit.

3.4.2 CONCEPT TESTING

If your feasibility and screening studies show that a product has the potential to be launched, the next natural step is to create it and test market it. It is not wise to purchase equipment at this time.

Packaging and labeling materials, as well as advertising and ingredients, should be the primary costs. Pilot plants are used by large corporations to produce small quantities of novel food items for testing purposes. Pilot plants at various area colleges or community centers can be used for free by start-up enterprises. Conversely, the test product might be made using suitable equipment at a nearby certified food processing plant (Kluytmans, 2010).

Consumer testing are sometimes undertaken as in-home usage tests at this point. Prior to the company starting a wider marketing campaign, consumers evaluate the product's likes and dislikes. Market testing is most effective when it is organized ahead of time with the assistance of a professional in the industry. Inquire about marketing experts who can help you create a strategy and analyze the outcomes of your test. Processing, formulation, and packaging should all be addressed in test marketing.

Figure 3.9: Concept testing is a great way for a firm to figure out if its product, design, service, or marketing is feasible in the target market and go on with the proper thoughts or ideas. Furthermore, concept testing will identify opportunities and dangers that will benefit from being addressed at the appropriate moment.

Source: https://www.marketing91.com/concept-testing/

A finalized formula is no longer a "recipe" at the time of product testing and should be presented in weight percent basis. All components should be sourced from several sources. These must be of excellent quality, with minimal variation between shipments. The method must be sufficient to produce a high-quality, safe product. Check for state or federal laws on processing conditions for certain items, such as final internal temperature. Packaging should appeal to the buyer while also protecting the product from infection. Using code packaging to keep track of shelf life and distribution might be beneficial (Khan et al., 2013). Documentation will be crucial in determining if your market test was a success or a failure. All processing methods and controls, including raw ingredient quality and temperature, ultimate cooking temperature, weight of each ingredient used in the physical, chemical, and batch test, chemical and physical tests done on net content of jars, and the number of faulty units, should be documented.

It's essential to keep the distribution region for test marketing as small as possible. By now, you should have a clear idea of who your target market is. Consumers should be given questionnaires to evaluate the quality of your goods. Maintain contact with the shop managers who sell your product, and make frequent visits to see who is buying it and where it is exhibited. Keep a careful record of the marketing exercise and seek assistance with data analysis to evaluate whether you should go to the next important step—commercialization.

3.5 MARKET STRATEGY AND BUSINESS ANALYSIS

If your market test is a success, your product will be ready to go to market. The product can still be manufactured at an existing food processing factory; otherwise, finding a place to produce the product is the key problem at this stage. A company must handle difficulties such as locating a building, location, utilities, equipment, location, and employees when establishing a processing plant. Consumer concerns should be taken into account during test marketing, and a second test may be undertaken if necessary.

The marketing of a product should be an inherent aspect of its commercialization. Companies having the financial clout to undertake nationwide marketing campaigns can choose a variety of approaches to get their essential components. National television, savings coupons, on ads, online advertisements, and product placement techniques are all common ways to sell new items. The promotion techniques are determined by the product and the target market (Joseph & Turner, 2020).

Finally, product management should be considered as part of the commercialization process. It should focus on improving quality and increasing profits. The product's quality is maintained by recording possible flaws as it is handled during production, distribution, and presentation. Cutting costs rather than increasing the price of your goods can help you increase profits without scaring away potential customers. To increase earnings, you must look for ways to enhance process efficiency, cut labor expenses, and identify alternative ingredient sources. Solicit consumer feedback while the product is still fresh to help find alternate flavors and packaging. 'Market strategies,' according to strategic management literature, are deliberate patterns of activities done in the market environment with the objective of increasing business productivity (i.e., optimizing shareholder returns and profits), such as pricing or mergers and acquisitions. Effective implementation of market strategies, according to Michael Porter, the originator of contemporary market strategy, necessitates a grasp of the competitive architecture of the sector in issue. For example, in a competitive business, a company that is unable to compete on pricing may opt to target a new consumer category by tailoring the attributes of its service or product to the target demographic.

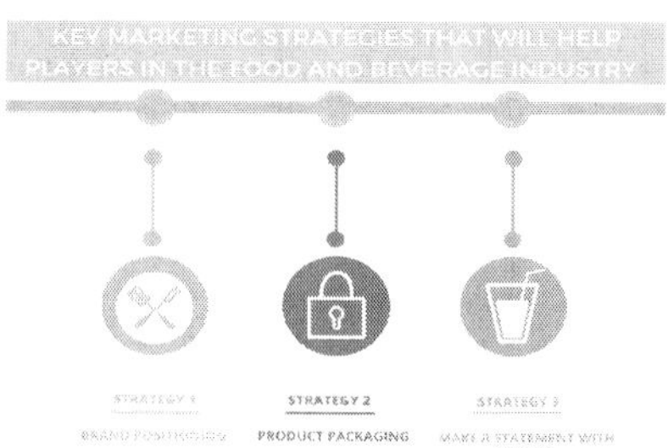

Figure 3.10: To stay afloat in this fiercely competitive business, food and beverage companies must step up their game. The marketing plan will emphasize the company's and goods' strengths. Implementing effective marketing techniques is the best approach for businesses to capture customers' attention and convert them into brand loyalists. For food and beverage sector companies to follow, Infiniti has compiled a list of four popular marketing techniques.

Source: https://www.businesswire.com/news/home/20180227005853/en/Four-Dominant-Marketing-Strategies-for-Food-and-Beverage-Industry-Players-to-Follow-Infiniti-Research

To limit the danger of losing market share, an incumbent business may opt to purchase new enterprises that enter the market. According to Porter, a firm's choice of acceptable market strategies includes examining how gaps in an industry's competitiveness might be exploited to create profits. This viewpoint emphasizes the connections between the implementation of effective market tactics and the possible need for government action to reduce or prevent the employment of anti-competitive techniques (Jankovic et al., 2010).

Market tactics, it's worth noting, work in concert with non-market initiatives. Non-market strategies are a collection of tactics aimed at improving overall business performance by influencing the interrelated policy, regulatory, institutional, ideological, and larger socio-political structures that define market settings. The difference between non-market and market strategies is often false, since both are sometimes compressed into single acts or activities carried out by businesses. For example, corporate social responsibility programs are frequently seen as having both a market-strategy and a non-market-strategy dimension (e.g., through improving brand value).

As a result, market strategy analysis is frequently inextricably linked to non-market strategy analysis, as the two can be mutually interconnected and overlapping. However, evaluating market tactics separately and entirely as a heuristic device is significant and crucial, particularly for understanding how market-based power disparities are formed or aggravated, and, extensively, for understanding corporate power and behavior.

We concentrate on the marketing methods utilized by processed food and beverage industries with significant market strength (hereinafter dominant food companies). Many of the world's most powerful food firms are global in nature, at least in terms of production and consumption. However, there are also instances where a single company dominates a single national market. Furthermore, markets have geographical borders, and regulatory authorities often perceive market power as being constrained inside these bounds (Bellows & Anderson, 2006).

The majority of dominant food corporations make and market branded processed food items, which are defined as enterprises that convert post-harvest agricultural commodities (together with intermediate inputs from main production firms and/or synthetic food component producers) into processed foodstuffs for the purposes of this article. There are two types of processors: primary and secondary.

Processing basic agricultural commodities, such as crushing oilseed, refining grains, or slaughtering animals for meat packaging, is how primary processors add value. Bunge, Daniels Midland, and JBS, are just a few of the well-known companies that engage in primary processing. Secondary processors, on the other hand, manufacture highly processed meals to add value to processed components such as refined sugar or corn syrup.

The Coca-Cola Company, the world's biggest soft drink maker, and nestle, the world's largest food manufacturer, both engage in secondary processing operations. Nestle engages in a variety of processed food industries, including ice cream, confectionery, and processed milk products. It's worth noting that many major companies work at many stages of the food value chain, and many food producers do both primary and secondary food processing.

The processed food manufacturing industry is extremely oligopolistic in terms of market structure (i.e., dominated by just a few businesses) and characterized by a large number of suppliers who can only sell their products to a limited number of major food corporations in numerous food sectors throughout the world (i.e., these constituting an oligopsony). Furthermore, processed food industries often have high entry barriers, allowing existing corporations to entrench their market clout even further.

The capacity of major food firms to reinforce and promote food environments and food value chains tailored toward the manufacture and consumption of their goods, many of which are highly processed, is a public health problem. To reinforce the justification for focusing on prominent processed food firms, we use the NOVA food classification system. The NOVA system, which is becoming more widely used in public health nutrition research, divides food products into four classes based on their degree of processing. Unprocessed and lightly processed foods make up the first group (Hirschi, 2012).

Processed culinary components, which include lard, sugar, butter, oils, and salt, make up the second group. Processed foods, on the other hand, are manufactured by adding sugar, oil, salt, or other ingredients from the second group to meals from the first. Ultra-processed foods (UPFs) are the fourth and final group, which includes soft beverages, confectionary, sweet biscuits, ice cream, and savory snacks. UPFs are created by mixing foodstuffs with synthetic additives, usually through a combination of industrial procedures and processes.

Figure 3.11: The ability to create good food is the first and most important need for sustaining a food and beverage business. You must know what your target consumer wants, or build the desire for your product offers, whether you are a chef or employing numerous chefs, or even creating your own bakery.

Source: https://www.designhill.com/design-blog/11-highly-effective-marketing-strategies-food-beverage-business/

Importantly, UPFs contain food ingredients that are never, or extremely seldom, utilized in residential kitchens, implying that they are nearly entirely manufactured by processed food makers. In highly developed market countries, UPFs currently account for over half of total dietary energy consumption, and their revenues in less developed market nations are fast expanding.

A second key public health effect of dominant food businesses' well-known concentrated market power is that they exploit their market dominance to produce profits that are higher than what would be possible in competitive marketplaces. This surplus of monetary resources can then be utilized to support corporate strategies and activities that harm public health, like lobbying, public relations, intensive marketing, and scientific institutions and research (Ilan et al., 2018).

Too frequently, food experts seek marketing solutions without first grasping the fundamentals of brand recognition. Before deciding on a route to follow as a small company food marketing specialist, you should concentrate on establishing and obtaining a footing in your target market. Some current basics of food marketing plan include:

- Creation of the brand awareness – A food architect's best buddy is Twitter. You may quickly contact your target clients via Twitter (and other social media channels). They can also contact you and provide comments or insights that previously took months to obtain. In-store questionnaires are still a wonderful method to connect with customers and build brand awareness. These surveys may be combined with culinary presentations to allow customers to test your items and provide feedback.
- Influencing of consumer preferences – The first stage is to raise brand recognition; after that, you'll need a strategy for influencing customer preferences. Set up a blind taste test in a neutral place where customers may try both your item and that of your rivals. Determine which brand they like and, more importantly, why. Is it because of the flavor, the presentation, or what they feel you represent? The creation of a product positioning statement is a prerequisite for considering your overall brand experience (Blissett & Fogel, 2013). Make it clear to the public whether your method is "green" or whether you support a certain charity. Any feature that can set your items apart from the competition will help to improve the perception and preference of your brand.
- Establishing and maintaining brand recognition – You want customers to recognize your brand and link it with pleasant feelings when they see your logo, tagline, or commercials. Consider advertising with a catchy melody that you can sing along to, even if you have heard it a hundred times. You are familiar with the jingle's product and the company that made it. This is what you desire to achieve and keep. You may try making a jingle that is both memorable and obnoxious, and then working to have it aired on the television, streaming media, and/or radio.
- Perceived value – Value judgments occur in a variety of forms and sizes. Quality, pricing, and quantity all influence value. Each customer will place a different value on different things. Perceived value refers to the consumer's perception of value. It's what they're looking for in a product. This might be container dependability, a longer lifespan, easier-to-read labeling, or being healthier while remaining economical. To have the best-perceived value, products and services do not have to be the cheapest.

- Life cycle of the product and the brand – May you be working on a new product release, a relaunch, a seasonal product, or are you the single brand that people have trusted for generations? Each of these may be positioned as a favorable brand benefit.
- Appropriate channels of distribution - If you are selling a youth-oriented product, on-the-go, or a single-serve, convenience shops are a must-visit. If your product, on the other hand, lends itself to bulk sales, target warehouse club retailers. Remember that conventional supermarkets now represent less than 50% of all food sales, thanks to the rise of e-commerce and newer technology. You should consider e-retailers as a source of distribution (Granato et al., 2010).
- Market share value - Even the most powerful businesses in a given field can lose market share. Because the pie's size is limited, the only way to obtain more is to grab it from someone else. Make a strategy for squeezing additional slices of the pie. You may consider line expansions (e.g., more goods, more SKUs), expanding into new sales channels, or forming a strategic joint venture with a non-competitive partner.

It's all too easy to slip into the trap of only advertising the weekly bargain. While this strategy may keep your items on the shelf, it offers little to help you develop a brand. Consumers are fickle, and they will consider switching to save money. You must constantly publicize and promote your brand's tangible and intangible features to generate loyalty. This endeavor is aided by participating in cause-related marketing. Customers will always support brands that support their ideals, lifestyles, and values. Customers will continue to buy from you and support your business if you show that you care for their beliefs.

3.6 PROTOTYPE DEVELOPING AND TESTING

Prototyping is an important step for both new and existing enterprises. Developing new goods is fun, but you will need a disciplined strategy if you need to meet specified nutrition, shelf life, health, or cost goals. Ingredients and their supply, shelf-life and stability, conformity, and claims are all important factors to consider. Getting too far along in the process without ensuring that the main requirements are completed might be expensive. As a result, many of our clients consider prototyping a food product to be an essential service (Granato et al., 2010).

A food prototype, described as a three-dimensional representation of your food product or concept, is an important first stage in the creation of novel foods, beverages, and nutraceuticals. Food Research Lab provides a wonderful opportunity to see your concept turn into something real and realistic, whether you are building a prototype based on established food products or inventing a new one.

Figure 3.12: A food prototype, described as a three-dimensional representation of your food product or concept, is an important first stage in the creation of novel foods, beverages, and nutraceuticals.

Source: https://studylib.net/doc/25456796/food-prototype-development---food-product-prototype---foo...

Creating a prototype, which is simply described as a three-dimensional manifestation of your concept, is an important early stage in the inventing process. Making a prototype may also be one of the most enjoyable and gratifying aspects of the process. This is because creating a prototype allows you to fully delve into your imagination, allowing you to leverage the talents that originally sparked your innovation concept. Seeing your concept develop into something actual and real, whether you are constructing your prototype at home or employing the services of a machinist, seamstress, or an engineer, is a thrilling experience.

Meeting with clients – either in person or electronically – to discuss the basic requirements. What would a person desire or need to accomplish? What market(s) are you aiming for? Which features of the product are required and which are optional? Having these documented and clear will assist you in ensuring that your product offers what you want (Gaudin, 2017).

There is a need to begin to offering recommendations regarding the product and how it will function based on expertise, experience, and study.

Descriptions on how it will offer the claims and advantages one requires while also balancing elements like as flavor, shelf-life, and nutrition. Mostly there is a normal advice of clients about any extra claims or perks that might be included in the formula. There is a need to put together an outline specification and nailing down the alternatives at this point.

Now it's time to write down everything that has been discussed and agreed on so that we can build a specification. There is a specification of the product, its materials, and the procedures that will be employed, as well as the outcome. One will get a breakdown of the components and their composition, as well as nutritional information and proof of the recipe's claims. They can tell you how long it will probably last and if they propose frozen, fresh, ambient, or chilled distribution. They will have done some patent and claim research as well.

To get to this point, prototype developers probably developed and tried a few prototypes, but now it's time to test and iterate the recipe. They may frequently incorporate a 'development day' with the customer in attendance, when they create tiny samples and variants. They examine the product's flavor, color, and texture, as well as customer comments, before making changes. Clients occasionally request consumer testing of a batch of items. This can be a crucial phase in the process of transitioning from a prototype to a finished, production-ready product. You may be going to a trade exhibition or a customer event. If that's the cases, they can give you with items to offer at tastings.

3.7 PRODUCT LAUNCH

If one is trying to produce a new food product to sell in supermarkets, convenience shops, or other retail outlets, here's how to test, validate, and expand it. Is this a little too simplistic? Sure. However, it is worthwhile to outline the road that many firms have traveled. It should be stated right away that there is no low-cost method to achieve this, and that launching a product requires some form of money (Dobrucka & Cierpiszewski, 2014). Whether self-financing or obtaining funds from a third party. It is vital to break into retail channels and have adequate marketing budgets to generate demand early on.

First and foremost, coming to market for a product is pointless unless you have done the essential market research. That is because merchants will constantly want to know if something will sell. Is it anything that people want? What categories of customers are likely to purchase it? So, getting some dependable and repeatable market testing done is the first step in launching a food product. This may be as simple as selling it at a street fair or a farmer's market. It might also be on consignment at a local store. And, ideally, you, the founder and product developer, or people close to the product's origins, should be meeting and welcoming consumers in person, as well as watching their reactions to it.

Figure 3.13: Developing a fantastic food and beverage product is only the beginning of your journey, and the prospect of taking on all of the chores by yourself may be too frightening.

Source: https://www.toastdesign.co.uk/food-branding-and-packaging/launch-a-new-food-product/

Once you are confident that your product has enough client interest, the next step in the expansion and validation process is to test it in a shop. To get your goods genuinely put on retail shelves. Because it is a more accurate test of a product that you may want to expand into retail outlets. To put it another way, the next phase of testing is to see where you'd sell it. From the back of the shelf. Product sampling in the shop and other sorts of promotion to increase sales can still support this type of in-store, on-shelf testing. After all, you are not yet a well-known brand (Dimkpa & Bindraban, 2017).

You are ready for a large roll-out if you've received validation and frequent re-orders for your product after a few businesses have consented to stock it on their shelves, where it will compete with other items. When it comes to a roll-out, you will need to get professional help (ahem, someone like me!). Now is the moment to get expert assistance in order to capture the attention of retail buyers on a broader scale and get your goods into several locations. For a regional shop, this might be for the entire chain, or simply one region for a national business. However, almost all grocery stores are localized, with no genuinely national chains present in every area.

In this world of expansion, you'll almost always need to hire a sales and marketing agency (SMA), which is also known as a broker, but SMA is now more generally used. It is vital to hire the proper broker with the correct relationships with shop buyers. Here's a link to a blog article that explains how food brokers and sales and marketing companies' function. Specialized consultants can assist you in locating the proper broker and negotiating the appropriate contract with them. Accessibility to sales data will be a factor in deciding whether or not to use a broker. This is because you will want to know the sales patterns of other items in your category as you build your product for following phases of your roll-out (da Silva Vaz et al., 2016).

With the proper team of advisers on board, including your sales and marketing agency, you will be able to take your product expansion to the next level, ideally establishing yourself as a strong and successful company. Whether you want to grow your business through cash flow, get more funding, or sell it to a larger firm, the team can help.

CHAPTER 4

THE IMPORTANCE OF NEW FOOD PRODUCT DEVELOPMENT

CONTENTS

4.1 INTRODUCTION

International competitiveness, rapid technological change, and shifting patterns of global market opportunities force companies to invest in New Product Development (NPD) on a regular basis, if not for financial gain, then for continued existence, and this is regarded as the key to success. New product development and advancement is widely recognized as a key generator of competitive advantage. Considering the significance of NPD for a company's current and future success, a large proportion of new products seriously struggle when they are introduced to the market (Costa & Jongen, 2006). Most novel idea constructs perform poorly into becoming commercially successful without the help of a structured process, according to research. As a result, formal NPD procedures have had a constructive impact on the management and control of some companies' new product programs. As a result, new products, if handled correctly, can provide a significant boost in growth that is typically unavailable from current models.

4.2 RISK OF FAILURE

Rates of return vary depending on the potential consequences of failure intrinsic in each new product situation. A main determinant in determining whether or not to move ahead with a new product is the combination of investments, risk, and returns. Because new product prediction models can be costly in terms of both time and finances, careful consideration must be given before implementing them. The impact of successes or failures on new products over a 5 years has been numerically demonstrated. It's clear that half of the companies polled were successful, with at least two of major new products launched in the last five years. Even so, just under two-thirds of those half mentioned such success with their innovative products. As a consequence, the average values for new product accomplishments for producers selling to both industrial and consumer markets were the same. As a result, each company's success rate indicates the ratio of all significant new products launched in the market during the preceding years.

Manufacturers serving consumer markets are more likely to achieve complete success or failure than those serving industrial markets. Companies at either end of these extreme levels will include, to a greater extent than the average, a number of companies that have launched only a few major new products. As a result, it's easy to see how a company that only introduces one or two major new products over a period of time is either extremely cautious or lacking in new product innovation expertise. Furthermore, the small range of products at risk rises the likelihood of overall success (or failure). This research looks at new products that have been introduced to the market (Chen et al., 2013).

Recognizing that nearly every new product will invariably hold some level of risk does not preclude endeavors to minimize that risk. Experience acquired from past NPD missteps can result the NPD team to rush the process, preventing the team from performing as a unit. This is entirely plausible because NPD absorbs a company's financial and human resources, with no assurance of stand outs.

4.3 THE NEED FOR STRUCTURED NPD PROCESSES

With fairly short product life cycles and a potential for increased product variety, NPD teams are under constant pressure to develop a broader and more diverse holdings of new product possibilities while also managing the risks involved in taking them from concept to launch. Following that, in simple terms, to reduce the risk of failure. Systematic screening, monitoring, and progression frameworks, like Cooper's stage gate approach, have been used to deal with this effectively and efficiently. Most of these concepts are not new; for example, cross-functional teamwork and coordination mechanisms were first discussed in the 1960s, and NASA's 'phased review process' has been used as a stage gate model. Despite the importance of NPD for a company's current and future success, a large proportion of new products struggle when they are launched in the market. According to research, the majority of ideas that join the NPD process underperform towards becoming commercial successes; in reality, only 14% of them do (Caporgno & Mathys, 2018). Industries have been reluctant to provide resources to progress the NPD operation because new product failure percentages are so significant, and the expenses involved with NPD are typically high. On the other hand, it could be made the argument that there is already a widespread belief in the 'new model of good practice' about integrating the various tools and techniques available.

Figure 4.1: Food Product Development.

Source: https://slideplayer.com/slide/12329881/

As a result, if properly managed, new products can provide a significant boost in growth that is rarely aligned by existing products. As a result, it's critical that businesses avoid the "Not Invented Here" (NIH) syndrome.

Defining and Measuring Success It is well understood that the success of a new product cannot be assessed in absolute numbers. It should be described and interpreted in light of realistic aims and priorities that reflect the unique circumstances of a new product. For several years, academic researchers have been studying the success (and failure) of new products. It's difficult to draw valid comparisons and conclusions because the various studies used various metrics to give feedback from a broad variety of industrial and market segments (Chaudhry et al., 2008).

As a result, determining common factors that result in new product success is difficult. However, two fundamental points can be drawn from the studies conducted:

- The success of a new product is highly dependent on the circumstances.
- There is no single factor that determines the success of a new product.

NPD success is highly situational, and there are only a few things that can be done to ensure it. As a result, companies establishing new products must regularly evaluate their own situation and recognize the many variables that can influence success.

4.4 CORNERSTONES OF SUCCESS

Recognizing new product critical factors has become a hot topic of debate among academics and business leaders alike. In order to gain a better comprehension of the launch of innovative products, it's important, if not critical, to ask what distinguishes success from failure and the rationale behind the factors. Process, Strategy, and Resources are the three central tenets of NPD success, according to research (Capitanio et al., 2010). At the business unit level, the three critical success factors were discovered to be the factors of new product performance. This study looked at ten different performance indicators, which were then summed up into the following points:

- Possessing a product innovation approach for the company that relates product design to the company strategy, identifies product development focus areas, has a longer-term drive, and is obviously communicated to all employees is sufficient to result the company to success.
- Adequate spending and resources, such as having the essential individuals and R&D spending in place, are another component that can add to a company's NPD success.
- Finally, and perhaps most importantly, having a high-quality new product process to direct product innovations from concept to launch is critical. It is the lesser praised of the three, but it has the greatest impact on the company's NPD performance.

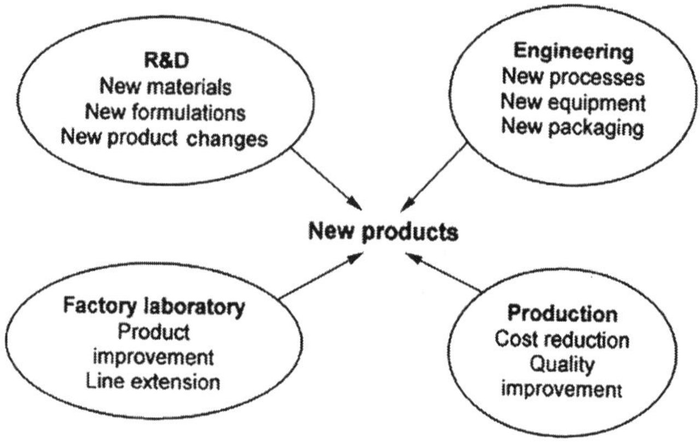

Figure 4.2: Technical product development.

Source: https://nzifst.org.nz/resources/foodproductdevelopment/Chapter-6-8-1.htm

Managers must consider the concept of a new product. The trouble of creating a new product confronts both new and established businesses. Some components of novelty in a product are desired by buyers. Production, marketing, location, service, use, and price are all examples of newness.

A new product is one that has new features or benefits. In terms of usage, visual appeal, quality, taste, price, and fabrication, it varies markedly from currently available products. A new product is one that consumers perceive to be novel (Yang et al., 2017).

- It could be a new brand, an improved product, or a customized product.
- A new product raises a number of issues, including how to develop it, launch it, manage it, why it fails, and what countermeasures should be brought to decrease its failure rate, among other things.

The term 'new product' is defined as follows:

- A new product is one that is provided for the first time to a specific group of buyers.
- A new product is one that is significantly different from existing products in terms of quality, features, or both.
- A product that is considered unique by customers is referred to as a new product. Consumers must be unfamiliar with it.

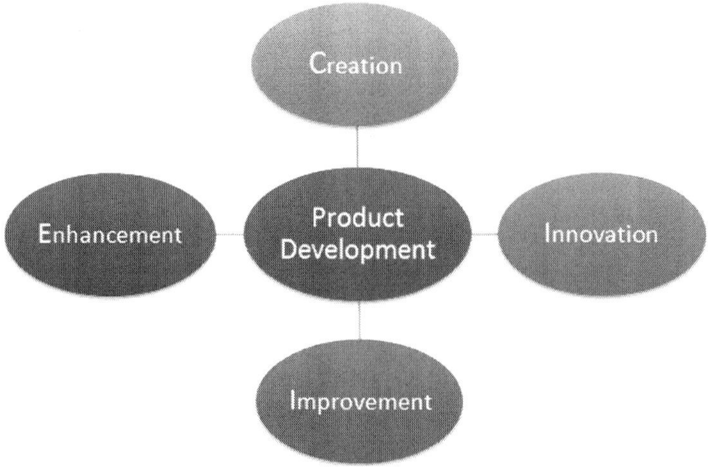

Figure 4.3: What is Product Development?

Source: https://kalyan-city.blogspot.com/2012/02/what-is-product-development-meaning.html

A new product is one that consumers perceive to be new. If we use this description, we may be able to find a variety of new products. Allen and Hamilton has identified six new product categories.

- New to the World: It is truly innovative, having been introduced to the world for the first time. Previously, the product was not available anywhere in the world. A new product could be pills to treat incurable diseases such as HIV/AIDS, diabetes, or a completely pollution-free vehicle (Wyrwa & Barska, 2017).
- New Product Line: When a new product line is presented in an established market, it is a new product. For example, Bajaj Auto Limited's introduction of a car to the market could be considered a new product.
- Addition to an Existing Product Line: It can be a new product if some models or styles are added to an existing product line. Hero Honda, for instance, introduced the CBZ and PASSION models in its two-wheeler product line as new models.
- Enhancement of an Existing Product: In this case, the existing product is enhanced with new features, qualities, or services. For example, the Pentium IV computer is an upgrade over the Pentium III.

- Repositioning: Repositioning entails launching an existing product into a new market or territory.
- Cost-cutting: Selling a high-priced products at lower prices or at a discounted rate results in the creation of a new product (Wenzl et al., 2006).

4.5 BASIC CONDITIONS/FEATURES OF A NEW PRODUCT

If a product meets the following criteria, it is considered new:

- It must appear to buyers as if it is brand new.
- It must stand out from the competition's products in a significant way.
- The customer must agree to it.
- It must be safe and easy to distribute (Vidović et al., 2022).
- It must be well-packaged, branded, and labeled.
- It must deliver satisfactory results.
- It must be generated within a cost range that is acceptable.

Figure 4.4: New Product.

Source: https://www.newswire.com/blog/how-to-use-a-press-release-to-promote-a-new-product-2

The following are some of the reasons for creating a new product:

A new product, also known as innovation, appears to be required due to a large number of factors. Because of the fast-paced nature of the business world, there are only two options: be innovative or die. Marketers must look for better ways to meet rising expectations (Abbas et al., 2010).

Companies that fail to recognize or react to the need for time must exit the market. The survival, growth, and development of a company are all directly affected by innovation or new product development. A marketer is always striving to provide a better product to his customers.

He aspires to provide higher levels of satisfaction than his competitors. He also tries to match products and services to the target market's changing needs and wants. A company's offer/product must change in response to changes in consumer fashion, personal taste, habits, and attitudes. The primary goal of new product development is to meet market demands. Even so, there are several other compelling reasons for a marketer to pursue new product development.

Figure 4.5: Food Product Development Process.

Source: https://discoverfoodtech.com/food-product-development-process/

4.6 MAIN FACTORS OR REASONS FOR DEVELOPING A NEW PRODUCT

4.6.1. Changes in Market

In comparison to the past, today's market is much more dynamic. Today's marketers face enormous challenges as a result of education and awareness, worldwide marketing, fierce competition, and the availability of a variety of substitutes. Market vogue, personal taste, and habits change on a regular basis, and marketers have no choice but to adapt to these changes by responding positively in terms of innovation. As a result, one of the most important drivers of innovation is consumer behavior.

4.6.2. Technological Changes

New production strategies are developed as a result of continuous technological advancement. Older technology and manufacturing methods are being phased out in favor of newer ones. A business invests a significant amount of money in technological research. New products are created to keep up with technological advancements (Arzani & Ashraf, 2017).

4.6.3. Increasing Competition:

Greater competition is one of the major factors driving people to seek out new ideas. Every business strives to attract and keep customers by providing superior products. Product innovation appears to be necessary in order to provide more comparative advantages and to satisfy customers more effectively and efficiently.

4.6.4. Risk Diversification

A company may develop new products solely to diversify risk. Current products may not have been able to meet market demands and expectations. A company can reduce the extent of irrelevancy by offering more options. As a result, the demand for continuous innovation grows as older products are phased out of the market.

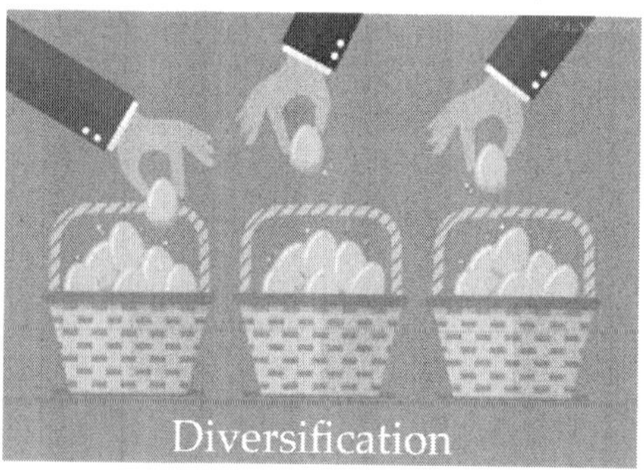

Figure 4.6: Diversification.

Source: https://www.iedunote.com/diversification

4.6.5. Image and Goodwill

Innovation is used to establish an image and reputation as a forward-thinking and dynamic company. The company wants to persuade the market that it is serious about meeting customer expectations. Obviously, a company that releases new products on a regular basis has a better reputation and can easily attract customers.

4.6.6. Excess Capacity Can Be Used

Excess capacity can be in the form of generating capacity or human skills. A company may decide to develop a new product in order to make the most of its plant and material capacity. Extra managerial or human capacity can sometimes tempt a company to pursue innovation (Vigani et al., 2015).

4.6.7. Seasonal Fluctuations

New products are sometimes created solely to reduce seasonal demand fluctuations. A business can achieve seasonal market demands by developing new products. Due to the corresponding products in each season, the market is satisfied, and the company can generate attractive business.

4.6.8. Growth and Development

Increasing market share or sales through innovation is a powerful strategy. Innovative products can help marketers take advantage of new opportunities. When existing products are unable to accelerate growth, a company is expected to create new products in order to expand its market, increase sales, and increase profits.

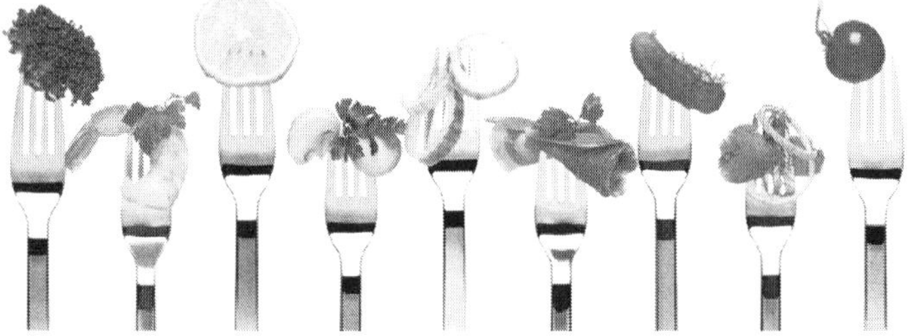

Figure 4.7: The Importance of Food Development.

Source: https://www.translatemedia.com/translation-blog/importance-food-language-development/

It takes more than perfecting a recipe to develop a food product. Producing new food products necessitates extensive planning, hard work, and extensive research. It is crucial to establish clear targets and time frames that integrate the business's future direction before beginning a new development venture. Companies develop new products in the hopes of gaining new customers, entering new geographic markets, maximizing revenue, increasing brand excitement, or expanding the market. Each year, thousands of new food products are introduced by both large and small businesses (Veeck, 2010).

For example, developing a line extension that uses existing equipment at a production facility takes less time than developing a new product that requires a specially made processing line. In some grocery categories, the failure rate of new products, which is defined as a product that is no longer on retail shelves after five years, can be as substantial as 90%. Food scientists, food engineers, regulatory specialists, marketing specialists, and buying gurus are all part of a larger company's product development team, whereas enterprises may not even have a research and development team. In order to create successful products, smaller businesses may rely heavily on outside resources such as universities and independent laboratories.

4.7 IDEA GENERATION

Companies use a variety of techniques to come up with new product ideas. Promotional teams may be tasked with the focal development of ideas, with researcher input supplemented. Consumer feedback may also provide inspiration. Some businesses may not require this step, particularly if they prosper on repurposing their competitors' products. A store brand item that is very comparable to a moniker brand product that is given is an example of this. Corporate idea generation can also include brainstorming sessions with participants from various departments. The major steps in creating a new food product can be divided into four stages after idea generation: screening, feasibility, product testing, and commercialization.

Figure 4.8: Best ideas on generation techniques.

Source: https://combustion.com/best-ideas-on-generation-techniques/

Attending trade shows, staying current with new product launches by other businesses, scanning scholarly articles and trade journals, and monitoring grocery shelves are all good ways to gather information about fashionable ingredients and consumer desires.

4.8 SCREENING

Following the creation of an idea, the stages of product design begin. The most important step in a development venture is screening. Thorough product concept testing can help a company decide whether to spend time and money in a venture or to abandon it entirely. Project suggestions should be in line with the organization's objectives. Throughout the development process, project managers should screen ideas to see if the market acceptability of the notion, ingredient availability, and regulatory environment have changed. Smaller businesses may enlist the help of outside firms to conduct market research (Van Huis, 2020).

During the screening stage, collaboration between departments aids in evaluating individual aspects of product development, such as financial and legal criteria, procedure and equipment accessibility, spending power and excipient accessibility, market shifts, and consumer perceptions. Product screening requires a thorough examination of markets and consumer research. When it comes to product evaluation, consumer testing is critical. Industries have no idea of verifying what consumers want, need, or are willing to buy unless they conduct consumer testing. Initial screening may expose information that will be useful in future marketing campaigns.

4.9 FEASIBILITY

Regulations, digital, and finances are all factors to consider when determining a company's feasibility. The tools will be accessible to respond initial questions of affordability that may be presented at any stage during the design phase by forming an interdepartmental team.

Regulations Firms must be aware of the state and/or federal attempts to regulate a product before beginning a project. State agencies regulate products sold locally (i.e., those that do not cross borders). Depending on the type of food, a product crossing state lines is under the jurisdiction of the US Food and Drug Administration (FDA) or the US Department of Agriculture (USDA).

Small food processing organizations are permitted in some states, but the handling area must be isolated from the living spaces by walls and doors, and there must be no direct access from the living area to the food manufacturing area. Because some local authorities actually prevent in-home industrial food processing, it's critical to be aware of local zoning laws.

Figure 4.9: How to Determine the Feasibility of a Business Idea?

Source: https://alcorfund.com/insight/how-to-determine-the-feasibility-of-a-business-idea/

"A raw, cooked, or processed edible substance, ice, beverage, or ingredient used or intended for use or sale in whole or in part for human consumption, or chewing gum" is defined as "a raw, cooked, or processed edible substance, ice, beverage, or ingredient used or intended for use or sale in whole or in part for human consumption, or chewing gum." Individual food product requirements are outlined in the code of federal regulations, which are referred to as standards of identity. In order to use a specific name, a product must meet certain requirements. To be labeled as "applesauce," for example, proper instructions on ingredient incorporation must be followed. If unsweetened, "applesauce" should have a soluble solids material of at least 9%, and 16.5 percent if sweeteners are added (as measured by a refractometer). Water, salt, apple juice, organic acids, nutritive carbohydrate sweeteners, spices, natural flavorings, and a color additive/color preserving agent can all be added in different amounts to the product (Tarasuk, 2008). The FDA and the USDA both release publications titled "Food Labeling Guide" and "The Food Policy and Labeling Guide," respectively, that lay out the guiding principles for labeling products that fall under their jurisdiction. The USDA Food Safety Inspection Service regulates meat and poultry products containing more than 3 percent fresh meat or at least 2 percent cooked poultry and are intended for interstate commerce (FSIS). All other food products are regulated by the FDA, excluding seafood, which is regulated by the Department of Commerce and the Food and Drug Administration. Heat-processed, low-acid canned foods, and acidified foods are subject to special regulations due to the risk of botulism and must be stored in hermetically sealed containers.

Acid foods have a pH of less than 4.6 and/or a moisture of less than 0.85. These rules are based on Clostridium botulinum and Staphylococcus aureus' microbial interaction. With the exception of alcoholic drinks, low-acid canned food is considered as processed foods with a pH greater than 4.6 and a moisture greater than 0.85. Water activity (aw) is a measurement of the amount of water available in a food for microbial growth. Low-acid foods that have had acid(s) or acid food(s) added to lower the pH to 4.6 or below with an AW greater than 0.85 are known as acidified foods. All food processors must complete an FDA-approved course of study known as the Better Process Control School.

In order for the FDA to approve a product, companies also need to provide specific processing information. Jams, marmalades, and barbecue condiments, for example, usually have pH values that are low enough to avoid these regulations. Low-acid food regulations apply to most canned vegetables and pickled foods. Other tightly controlled areas that need to be addressed fall into two categories: health and institutional safeguards (Sousa et al., 2008). Adulteration, organic toxicants, chemical additives, byproducts, and unsanitary preparation or retaining practices are all protected by health safeguards. Categorizing issues, particularly with regard to inaccurate and misleading declarations, and net materials are examples of economic safeguards.

4.10 TECHNOLOGY

The required parts, facilities, and processes to manufacture a commodity must be established before a new food product can be launched. When products are discovered to be technologically impractical, the project should be abandoned.

Figure 4.10: 2019 Best New Retail Products Contest.

Source: https://www.refrigeratedfrozenfood.com/articles/97699-best-new-retail-products-contest

4.11 FORMULATION

In order to find the best combination of ingredients, processing parameters, and packaging options to create the desired product, a variety of ingredients, process conditions, and packaging possibilities will be used. This step necessitates thorough statistical analysis and meticulous record-keeping. Following some preliminary testing, an experimental design will reduce the number of prototypes required, saving time and money. A laboratory notebook should be used to record all formulas and experiments. Each entry should include all pertinent information (Silva et al., 2020).

This is useful when projects are momentarily stalled, last for an extended period of time, or are passed on to other developers within the company. When choosing ingredients, think about whether the product will be available year-round or only during certain seasons. To test quality and cost effectiveness, product developers will typically consider multiple suppliers of the same product.

To meet their needs, bigger businesses may need to find multiple suppliers of the same product. If more than one provider is used, all suppliers must adhere to strict product specifications.

4.12 PROCESSING

What machinery how would you even have if you have an existing site? If at all possible, companies try to manufacture newly developed products on previously purchased equipment. Because purchasing new equipment is an expensive investment, many new product development projects focus on enlarging product lines using established facilities and equipment. As a result, product designers should be cognizant of the equipment available inside the location where the final product will be manufactured. If the product will be manufactured in multiple plants, the focus should be on the differences in equipment available and how they can be resolved to generate commercially comparable products (Siegrist, 2008).

4.13 FACILITIES

It is critical to consider the production plant that is available. If the company has bought the new plant, the water supply as well as sewage pipes will need to be examined. It's crucial to consider the plant's internal conditions, such as temperature and humidity regulation. In order to produce quality foods, facilities in areas with high heat and humidity during the summer months may need to change their operating conditions.

4.14 PACKAGING

Packaging plays an important role in a consumer's decision to buy a product, especially for first-time buyers. It's crucial to think about how customers will perceive the packaging and whether it will communicate the product's quality goals, such as whether it's a high-end premium product or a generic grade (Balasubramanian & Panigrahi, 2011).

Marketing, product designers, and packaging technicians should think about the packaging materials used on competitors' products and how they can differentiate themselves.

4.15 DISTRIBUTION

Frozen and refrigerated foods are examples of items that require special distribution. The expense of special distribution should be considered by organizations. The radius in which the item will be available is another factor to consider. Is the product going to be distributed nationally or regionally? The packaging requirements may be influenced by the distribution radius.

4.16 TIME ON THE SHELF

The shelf life of a product is determined by how long it retains its quality as customers perceive. When analyzing distribution channels, a product's shelf life is critical. Accelerated or real-time testing can be used to determine shelf life (Sanguansri & Augustin, 2006).

4.17 SAFETY

New product designers should think about their products' safety risks. The history of breakouts and accepted for publication safety risks associated with specific product categories can provide insight into risk factors that processors should be aware of. Peanut butter manufacturers, for example, must use control mechanisms to test for salmonella contamination following a large eruption in early 2009 that sickened over 400 people and resulted in at least five deaths. Some products are vulnerable to pathogens and spoilage microbial growth. Consumers are also at risk from allergens and physical contaminants such as metal shards from processing equipment.

4.18 FINANCES

Before a food product can be created for sale, it is necessary to have a comprehensive understanding of all manufacturing and sales costs. Prior to manufacturing, a thorough cost assessment should be performed. Fixed and variable costs are the two types of costs to consider. Yearly fixed costs are those that do not change from year to year, regardless of production levels. These costs include things like equipment, buildings, property taxes, and other things that aren't affected by production changes. Hired labor, raw materials, packing material, fuel, electricity, utility services, and other items for use during production are examples of variable costs. Before test marketing and commercialization, indirect expenses should be carefully examined to establish a unit price for the new product to make a profit.

4.19 TEST MARKETING

If your screening and feasibility tests show that a product has the potential to be launched, the next logical step is to develop it and test market it. It is not advisable to purchase equipment at this time. Packaging and labeling materials, as well as promotion and ingredients, should be the primary costs. Pilot plants are used by large corporations to produce small quantities of new food products for testing purposes (Rudolph, 2007). Pilot plants at many regional educational institutions or community clinics can be used for free by start-up companies. Put another way, the test product could be made with capable equipment at a nearby approved food processing facility. Consumer tests are sometimes conducted as in-home use tests at this stage.

Figure 4.11: Test Marketing.

Source: https://courses.lumenlearning.com/marketing-spring2016/chapter/reading-commercializing-new-products/

Prior to the company trying to launch a larger marketing campaign, consumers evaluate the product's likes and dislikes. Marketing research is most effective when it is planned ahead of time with the assistance of a professional in the field. Inquire about marketing experts who can help you create a strategy and evaluate the findings of your test. Formulation, preparation, and packaged foods should all be addressed in test marketing. A final equation is no longer a "recipe" at the time of product testing and should be articulated in weight percent basis. All components should be sourced from multiple sources. These must be of excellent quality, with minimal variation between shipments.

The process must be sufficient to produce a high-quality, safe product. Check for provincial or national regulations on process variables for specific products, such as final internal temperature. Packaging should appeal to the consumer while also protecting the product from contamination. Using code packaging to keep track of storage period and distribution can be beneficial (Roy et al., 2009). Documentation will be crucial in determining whether your market test was a success or a failure. All processing steps and controls, including raw ingredient quality and temperature, final cooking temperature, weight of each ingredient used in the batch, physical and chemical tests conducted on net material of containers, and the amount of defective units, should be documented. It's best to keep the distribution area for test marketing as small as possible. By now, you should have a clear idea of who your target market is. Consumers should be given questionnaires to assess the quality of your product. Maintain contact with the store managers who sell your product, and make frequent visits to see who is purchasing it and where it is displayed. Maintain a continuous record of the market test and seek assistance with data analysis to determine whether you should proceed to the next big step, which is commercialization

4.20 COMMERCIALIZATION

If your market test is a success, your product will be prepared to go to market. The commodity can still be manufactured at an existing food establishment; otherwise, finding a location to producc thc itcm is the primary concern at this stage. A company must address issues such as finding a location, constructing, machinery, utilities, and personnel when establishing a processing facility. Consumer concerns should be taken into account during test marketing, and a test series may be conducted if required (Roohinejad et al., 2017).

The promotion of a product should be an integrated component of its commercialization. Companies with the financial clout to fund national marketing strategies can take a variety of approaches to get their products recognized.

Savings vouchers, national tv ads, digital advertising, and commercial advertising strategies are all common ways to market new products. The promotion practices are determined by the product and the target market. Finally, product maintenance should be considered as part of the commercialization process. It should focus on improving quality and increasing profits. The product's quality is maintained by recognizing potential defects as it is handled during handling, distribution, and display. Cutting costs rather than having to raise the price of your product can help you increase profits without scaring away potential customers. To increase profits, you must look into ways of improving process efficiency, cut labor costs, and find alternative ingredient suppliers. Solicit consumer feedback while the product is still new to help identify additional flavors and packaging.

4.21 PRODUCT LIFE CYCLES

During the course of their sales, products go through cycles. Companies heavily market their brand during the initial rate when they are first launched in order to attract customers. Demonstrations in stores are often used to entice customers who would not otherwise try the product. Markdowns and coupons can also help a new product sell faster. The company's costs are high and the yields are minimal during this initial period. The next stage of the cycle is a period of rapid growth. Repeat customers may decide to buy the product on a routine basis at this point. Customers' recommendations may start to attract new customers (Reis et al., 2014).

Expansion into new markets may help boost sales. Costs are still high, but profits are increasing. The next stage is a decrease in the rate of growth. Repeat customers are decreasing, new markets are being tapped out, competition is increasing, and there are new costs involved with attempting to draw attention to the product. In this stage, profits are still good. Due to consumer fatigue, sales do not increase during the stability period. The product generates little enthusiasm, and sales remain stagnant. Profits and costs are equal, but profits may start to decline. Competitive products begin to outperform the product during the product decline phase, and advertisements are too costly to be advantageous. The product's sales are declining, and it's becoming more expensive to maintain. The item is not profitable. Companies must now decide whether or not it is essential to eliminate producing the product.

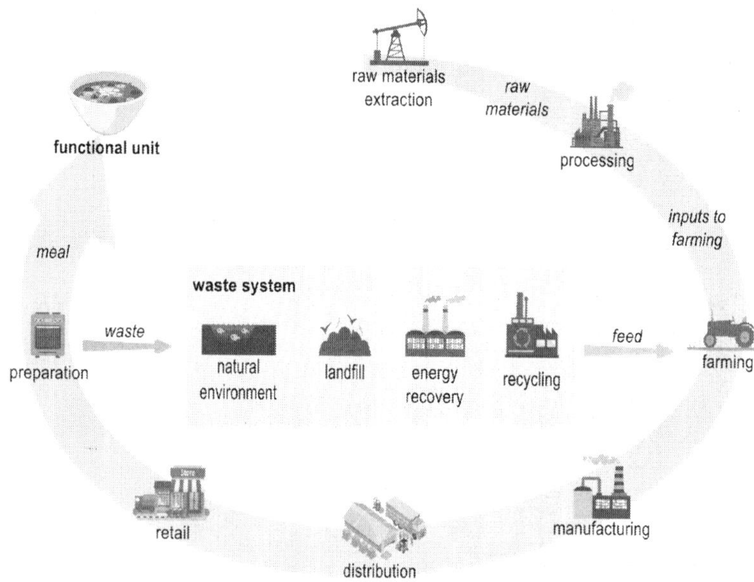

Figure 4.12: Life Cycle Assessment of Food Systems.

Source: https://www.sciencedirect.com/science/article/pii/S2590332219301289

New product development (NPD) is critical to a company's success and expansion. Standard models have been introduced in the high-to-medium-technology manufacturing sectors. The implementation of these systematic NPD procedures, aided by appropriate decision support tools, has resulted in significant cost savings, improved product quality, and supply chain availability. However, due to consumer spending for fresh and natural diets, particularly more nutrient dense low-calorie food, and a preference for highly customized and personalized food products, the challenges associated with food NPD are rapidly increasing. As a result, there are a plethora of new food varieties, types, and shapes being introduced on a regular basis (Pereira & Vicente, 2010).

The majority of these new products are developed in small to medium enterprises, which account for the majority of food manufacturers in most advanced nations, using company-specific incremental NPD procedures. This emphasizes the importance of more research into new NPD methodologies and technologies in the food industry.

While the food industry is often associated with a lack of research and technology adaptation, several emerging trends, such as increased consumer demand for fresh fruit and vegetables, natural foods, greater track and trace information, and an objective is to minimize meat consumption and increasing animal welfare, may change this. Meeting these demands will necessitate more than expanding supply; it will necessitate innovation in the shapes in which these foods are delivered to ensure that today's consumers have access to the widest range of flavors, colors, and nutritional value possible.

Large retailers' fierce competition drives innovation, which is then passed on to their manufacturers, leading to more efficient food packaging to extend shelf life and reduce waste, the use of alternative sources of protein, and product restatement subject to seasonal and locally sourced ingredients, for instance. The process of designing, manufacturing, and bringing a new product to market is known as new product development (NPD).

The majority of new food products, especially those made by small and medium-sized businesses, are created using company-specific NPD processes or models adjusted from other industries. Food production variables like consumer health, affordability and availability to ingredients, food supply chain strength, ingredient sustainability, and, more broadly, management of the food product's lifecycle must all be factored consistently, which is a major problem (Peters et al., 2016). There is potential for a formalized, food-specific NPD process. To imagine how such a process might work, one must first develop a deep understanding of the most common NPD models, as well as how they have been implemented. Then there are the unique challenges that the food manufacturing industry faces.

This brief communication presents an overview of the NPD process, discuss its significance and importance in the food industry, cite examples of NPD processes in industrial manufacturing, showcase their flaws, and define key aspects to consider in food product NPD. The process of converting a new market chance into a commercial entity through a series of activities with the goal of meeting specific objectives has been defined as new product development. So because effectiveness of a new product is dependent on a number of factors, constructing new products can be a difficult and risky business activity, but it is necessary for product market growth and competitive advantage.

Food products that effectively enter the food market substantially raise the value, and thus the price, of their ingredients, making NPD particularly important in the food sector. The commercial benefit of food ingredients at the farm gate accounts for only 20% of the commercial benefit of the final packaged food product on the store shelf for many processed foods. This means that domestic recipes and processing, which are defined as part of the NPD process, are critical to increasing the financial value of food products. Three kinds of knowledge are required to implement an effective NPD process. Understanding internal strengths and weaknesses, branding strategy, company strategy, competitors, distribution channels, and the larger market environment is the first step (Pereira & Vicente, 2010).

The second requirement is an understanding of the technologies required to manufacture the product. Finally, it's critical to understand current consumer demands and how they might react to burgeoning societal changes, and also latent needs that aren't being met by current offerings. NPD must be fully integrated throughout a company, from supervisors to manufacturing line operators, in order to be successful.

With a lifecycle view, all actors in an efficient NPD process must understand not only their particular project tasks, but also the probable downstream and upstream repercussions, like possible competitor responses and repercussions for raw material supply. To develop new products, food manufacturers employ a variety of approaches, methods, and techniques. Convenience product manufacturers, for example, frequently use Stage-Gate or milestone-based models, which generally include an idea scoping stage, accompanied by business case progress, product testing, evaluation, and launch. Other frameworks take a more holistic approach, which is frequently circular. The "PD Process Unified Model," for instance, deems the NPD process as a three-stage holistic approach that incorporates marketing and engineering.

The first stage, pre-development, involves evaluating market needs, core competencies, potential product lines that could meet these needs, and finally, affiliated business models. The second stage is development, which includes prototype creation, testing, and refinement. The third stage is post-development, which includes product launch, market performance, and feedback for future NPD. The various NPD models used are primarily based on various phases grouped around the above-mentioned pre-development, development, and post-development phases (Batista et al., 2013).

None of these designs were made solely for the food industry, which has its own set of characteristics. Ingredients used in food manufacturing, for example, are fresh and must be preserved under strict conditions before being consumed. Furthermore, even after a new food item has been crafted and manufactured, food manufacturers face ongoing changes in their production processes, such as ingredient sourcing, new demands to improve food safety, and shelf life stability, to name a few. Food producers may not have consistent product turnaround and secure supply chains because NPD models were not developed exclusively for food manufacturing. To better serve the food industry, NPD models must be modified for new food product development (NFPD). The following important areas should be covered:

- Meeting customer needs and sensory quality: Modern consumer demands emphasize quick preparation times, seasonal colors, and palatable textures.
- Seasonality of foods: Customers value seasonal ingredients because they believe they are of higher quality and are sourced locally. Given the limited accessibility during certain times of the year, relying too heavily on in-season foods can be dangerous.
- Local ingredient sourcing and distribution: There are numerous potential benefits to promoting community networks instead of relying on global supply chain. Reduced emissions from transportation could be one of the environmental benefits. Consumption of locally produced goods can help to create employment and reduce shipping costs (Barreteau et al., 2006). Finally, local ingredient sourcing can potentially boost the food system, attempting to make it more resilient to interruptions in certain aspects of operation while also potentially expanding the range of ingredients and thus nutritional diversity offered to people.
- Food hygiene is one of the most pressing concerns for food manufacturers. Food developers must understand the raw material and final product quality and storage requirements, as well as the product, formulation, and processes required to safely produce the final food product. As a result, from the start of the NFPD process, food manufacturers play a critical role in ensuring product safety and recognizing potential hazards.

- Ingredient and final product traceability: Food products and food distribution networks have evolved dramatically over the last hundred years, from the production of localized and seasonal food items to the production of sophisticated food products containing a wide range of ingredients from around the world. More recently, technological developments have enabled information about a food product's ingredients to be saved and communicated at all phases of the supply chain channel, improving monitoring. Internal and distributor traceability has been put in place as part of the quality control systems of huge corporations like Nestle and Walmart, and there are indications that most large manufacturers and distributors in industrialized states are following suit (Bellows & Anderson, 2006).
- Ramping up food recipes created by the NFPD team to an industrial level necessitates a number of steps, including adjusting process times and temperatures, considering hygiene circumstances in the manufacturing plant, determining logistics plans, including storage, taking into account the food product's characteristics, and selecting equipment and its working conditions.
- Environmental impact: Over the course of their lives, food products have a significant environmental impact. Food consumption accounts for between 15% and 28% of total emissions in advanced nations, as well as around 30% of energy demand worldwide.

Because it is guesstimated that around 80% of a product's environmental impact is determined during the design phase, NFPD can help mitigate these environmental impacts.

Over the years, the food manufacturing industry has successfully adapted numerous generic NPD style to meet its needs. However, changing consumer demands, intense retailer pressure, and growing pressure to reduce negative environmental impacts necessitate overhauling existing models (Blissett & Fogel, 2013). Any formalized NPD model designed exclusively for food manufacturers must include variables like closely customer's needs and expectations, using local and seasonal foods, guaranteeing food safety and greater transparency, guaranteeing production at scale, and reducing lifecycle environmental effects.

4.22 FOOD PRODUCT DEVELOPMENT

Small and medium-sized businesses (SMEs) played a key role in the development of radical innovations. The company should assess its actual conditions and either implement new methods or improve existing ones. Food product management is a dynamic process that producers are concerned about because consumers' needs change over time but they may no longer want the goods that were in high demand yesterday. Small businesses can thrive in the long run by making prompt planning and operational changes and encouraging close relationships between core shareholders and customers. SME should establish food quality and sensory benchmarks in the development of food products and include them in the requirements for final product QC.

It should include the findings of consumer preference surveys. Distributors, academic institutions, universities, affiliated companies, and customers can all contribute inputs and knowhow to the food product development cycle in SMEs. In SMEs, product development and innovation are ongoing processes driven by small business owners and their employees. Innovative products, new systems, new markets, and management concepts are all examples of these types of activities, and they all contribute to economic growth (PF Guiné et al., 2016).

New food products frequently fail since they're not tailored to meet the needs of consumers or are not manufactured efficiently. To design and manage product development projects, SMEs require qualified personnel or adequate support services, particularly if larger changes in the manufacturing process are involved. External acumen from research facilities, universities, and customers could be used to solve this problem. There are numerous reasons for the food industry's low rate of innovation. During the development process, the qualitative user prerequisites should be interpreted into product design parameters. Parameters are the food product's measurable characteristics and quality determinants.

One method demonstrated a method for clustering consumer feature preferences and transforming spontaneously written text about a specific food product into directions for food product development work. Firms require critical resources, including a substantial sufficient market, an acknowledged product or service, buying customers, and an appreciative network, in way to attain a stage of financially viable development. Furthermore, the firm must be able to effectively utilize these resources, i.e., through its motivation, competence, and organization. Each of the aforementioned factors plays a role in corresponding consumer needs and has an impact on the success or failure of food products.

Quality of food product as competitive advantage

Food product development is difficult for SMEs due to a lack of capacity, human resources, and financial resources, but consumers demand and expect safety and quality from food products on the market. Companies have well-defined and well-established quality and regulatory programs in place to pursue and maintain product quality, as well as to avoid negative influence on product quality, consumer complaints, and product rejection (O'Shea et al., 2012). The focus is primarily on constructive and chemical analyses; while many businesses have quality control sensory systems, only a few have effective quality control sensory systems. Enterprise management and policymakers should encourage the development and implementation of sensory programs within their organizations, and use them as a means of assuring high and consistent final product quality and marketable food manufacturing.

Some research offers a structured approach for classifying food industry products, understanding food industry characteristics, and understanding the food industry market structure. SME survival and success are likely if producers can identify and meet current and future consumer needs. Intake of goods for the purpose of displaying status is known as status consumption. It could be used as a quality criterion for product development, because modern status direction has a positive effect on consumer attitude and overall life satisfaction. The profitability of a company and the sale of a product determine the product's success.

4.23 THE BUSINESS JUSTIFICATION FOR PRODUCT DEVELOPMENT

Small manufacturing companies are facing extremely difficult trading conditions. It's difficult enough to get orders in, ship products out, chase payments, and maintain cash flow on a daily basis. Meanwhile, globalization and lower labor costs in other parts of the world have added to the pressures. Some local producers have reacted by becoming simpler and cutting costs, which is a great start. However, only one business can be the least expensive option in any given sector (Nicklaus, 2011).

Nonetheless, the ideal small businesses have had to reinvent in order to truly compete, fight back, and ultimately thrive. They've decided to do things differently and superior. Creating new products allows you to expand your market, gain market share, sell more, and boost revenue streams.

Meanwhile, redesigning existing products allows for cost savings, higher margins, and ultimately higher profits.

Aside from the commercial aspect, creating great products is extremely rewarding and something that most design teams strive for; goods that are appealing, dependable, and intuitive and user-friendly. Having great ideas is one thing; having an effective operation to produce a better product exactly as you envision it is quite another.

Small businesses that perform well plan and manage their product development operations ahead of time. They've created product development processes that are tailored to their requirements, and they're employing techniques that add real value. They're paying attention to what buyers want in terms of features, new products, and free customer service. This also entails soliciting feedback and ideas from anyone who interacts with the product at any point during its life cycle (Nedovic et al., 2011).

To develop and master their product development processes, the finest small manufacturing businesses collaborate and work with a variety of individuals and organizations. This is all in line with the company's overall goals. It should be mentioned that product development is not without risk. If done incorrectly, it can have serious ramifications for a small producer. The good news is that small manufacturing businesses have a number of natural advantages that increase their chances of success.

They have flatter organizational structures and can make quicker product development decisions as a result. They are also frequently closer to their final consumers, allowing them to obtain feedback more quickly and thus be more responsive.

The concern is how to get nearer to what the strongest are doing, instead of what the rest are still doing, which is again veering or floundering due to a lack of understanding of organized product development.

Why is it so important for product development to be precisely tailored to YOUR requirements?

Small manufacturers must get products to market quickly in order to stay competitive, thrive, and prosper. They must also move quickly through the development process. A great deal has been scribed about product development, and diagrams depicting a series of steps with various levels and review gates are common (Capitanio et al., 2010).

Even so, the issue with this is that it can appear long-winded and slow at first.

If this is the assumption workers have, the process is unlikely to be used, as the strain to get products out by a certain date often takes precedence. Crucial tasks, on the other hand, may be overlooked or rushed.

Any procedure must take into account your company's goals and circumstances. What's your product development goals and when do you want to achieve them? What role does this play in the overall business strategy? This final point is extremely significant. Each and every product development plan must be in line with the company's overall strategy. For starters, it will require the support of senior management, as well as resources.

Because your situation is unique, these considerations must be taken into account. Rather than being a generic process provided by a slick consultant, the product development procedures should be customized to achieve these goals.

What designers and technicians really need is a solid, flexible, and relevant process that can be followed and result in successful products. Furthermore, the product development process must be quick, with activities running simultaneously whenever possible (Monteiro et al., 2010). Groups from across the company should work together to achieve a common goal, enlisting outside assistance when necessary. A deep sense of management, combined with discipline, aids the process in moving forward and running smoothly.

Taste is the most essential factor for consumers when deciding whether or not they like a product. Every survey conducted by the International Food Information Council Foundation to date has validated this, with 88 percent of those polled citing taste as the main reason for purchasing a product in 2020. With figures like these, it's evident that taste should be a priority all through product development and a deciding factor in whether or not to commercialize a product.

Figure 4.13: Determinants of success.

Source: https://www.taste-institute.com/en/resources/blog/importance-of-taste-in-product-development

The "first tipping point" in product development, which takes place when a customer chooses to buy a new product, has received a lot of attention. A variety of factors may affect their decision; for example, the packaging must appeal to them, the price must be reasonable, or they wish to try a new and innovative product. A high trial rate can be achieved if a significant number of individuals are interested in your product.

While a high trial rate is encouraging, it is not a consistent indicator of a positive outcome and may only reflect the marketing strategy's effectiveness. The true measure of a product's success is determined by the "2nd moment of truth," or when consumers currently consume it. If they have a positive experience with a product, they are much more likely to buy it again. As a result, there is a high percentage of repeat customers and a successful product (Morgan & Sonnino, 2010).

Taste is the most important repurchase driver for food and beverage products, and it is central to the second moment of truth. "When you see a latest brand that asserts to be natural, all organic, without preservatives, et cetera, you could perhaps buy it because of these claims," says Jason Cohen of Analytical Flavour Systems. That's how marketing works. However, you will only purchase it if you enjoy the flavor."

Figure 4.14: Food taste stands supreme.

Source: https://www.taste-institute.com/en/resources/blog/importance-of-taste-in-product-development

4.24 THE IMPORTANCE OF LOYALTY

Repeat purchases create loyal customers who are critical to your product's success. According to a study by Löfgren et al., companies can increase their profits by nearly 100% by retaining just 5% more of their customers, implying that providing customers with a great-tasting product which they will want to buy again is highly profitable (Mitterer-Daltoé et al., 2014).

Customers who are loyal are willing to market to, more likely to spread the message about a great product, and more responsive to other products from the same brand. According to statistics, it takes eight purchases for customers to become loyal, and between the first procurement activity and the first repurchase intention, a brand loses about half of its customers. It will lose half of the remaining buyers from the first repurchase intention to the second. This cycle will continue until it reaches a point of stability, which will be around the 8th purchase.

According to Nielsen, brands will not make it to the eighth purchase without a strong product experience, particularly given the intense contest and inevitable variety-seeking that is component of the contemporary consumer experience. Because taste is the most important factor in deciding whether or not to buy a food or drink product, having a great taste is essential for creating a positive product encounter that will inspire customers to purchase your product again and again.

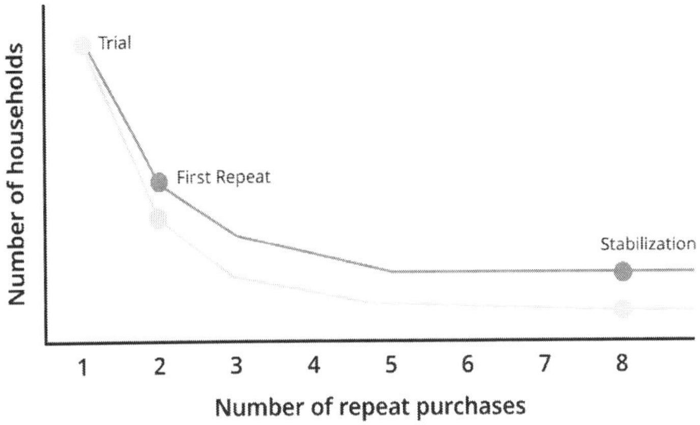

Figure 4.15: Repeat purchases statistics.

Source: https://www.taste-institute.com/en/resources/blog/importance-of-taste-in-product-development

4.25 BAD TASTE CAUSES PRODUCTS TO FAIL

However there are a number of factors that can influence a new product's success or failure (e.g., speed to market, sales prices, packaging, etc.), taste reigns supreme in the food and beverage industry.

Even the safety of a well-known brand can't save a product if consumers don't like it. When a well-known US beverage company altered the formula of their product, they discovered this the hard way. Only 13% of customers liked the new product's taste, according to polls, and the employer gets thousands of formal complaints (Matser et al., 2004). After only 79 days on the market, the product was removed and the initial formula was reintroduced. "Compromising on taste not only decreases the chances of a product's success, but it also puts extra stress on all the other influencing factors to grab for an average product," says John Stanton of Food Processing. In many cases, this results in a lower price to entice people to purchase the product."

Consumers, on the other hand, are willing to pay extra for high-quality product lines, and a flavorful, well-developed item can significantly boost your profit margins.

4.26 CONCLUSIONS

Producers who identify and satisfy present and future consumer needs can expect SMEs to succeed and survive on the market. The gathering of relevant information about consumers' needs and expectations is a prerequisite for consumer-oriented product development. Sensory information about food product quality and consumer preferences can be obtained through consumer sensory tests, which is valuable and helpful for quality assurance and quality control. Enterprise management and policymakers should encourage the development and implementation of sensory programs within their organizations, and use them as a tool to ensure product quality and competitive food production. Consumer-focused food product development should be viewed as a tool for gaining a competitive advantage and long-term market success, as well as for avoiding negative changes in the product quality and appropriateness, as well as consumer complaints and product rejection (Chaudhry et al., 2008).

The success of a product is determined by a variety of factors. A product with realistic goals and prudent fiscal analysis is more likely to succeed. Collecting sufficient product research aids in the development of products that are both consumer-friendly and competitive in the global market. To build a realistic, profitable business/product, you'll need a good business model with sufficient lists of all tools needed. User acceptance, timing, and luck all play a role in product development.

CHAPTER 5

CHALLENGES OF NEW FOOD PRODUCT DEVELOPMENT

CONTENTS

5.1 INTRODUCTION

There is a large market potential and cap in the food and drink industry. The industry also has intense competition. Existing products have to be innovated constantly to ensure that a bran remains relevant. Relevance can also be achieved through the launch of completely new food products at affordable prices. By launching and developing new products, brands are able to retain their consumers, stay ahead of competitors and gather new customers (Martelli et al., 2020). However, developing new products is not easy. It is often at times accompanied by resource requirements and huge costs. Currently, it is estimated that about 60% of the food manufacturing companies are willing to invest in new food products. The number is expected to increase in the coming years. For this reason, food technologists aspire to venture into food and beverage consultancy.

Figure 5.1: The food and beverage industry is ever growing.

Source: https://www.google.com/amp/s/www.consultancy.uk/news/amp/22506/five-top-trends-impacting-the-food-and-beverage-industries

Some of the challenges faced on the production of new food products are discussed as follows:

5.2 DEVELOPMENT OF NEW FOOD PRODUCTS REQUIRES SKILLS

Skills are very useful in ensuring that money provided by the company is placed into good use. This means that access to such skills places funds used at stake. The creation of both old and new products is not an easy task. The same applies to the regulation of such products. There are several factors that much come into place to ensure a successful production. The safety and quality of a product can be affected by a bad thermal process, microbial concerns or problems with ingredients (Lusk & Briggeman, 2009). Most of the skills are obtained through education in higher learning institutions or tertiary institutions. Individuals on the food production industry need to possess the ability to determine the opportunities and challenges in their industry. The opportunities include the creation of successful and delicious food products that can be sold in markets and groceries while the challenges include the acceptability of the product on the market and ensuring profitability of the product.

Figure 5.2: Skills in food production allows creation of new food products.

Source: https://qualdes.com.au/.well-known/captcha/

Food developers should be able to keep their ears on the ground and speak with people, read food blogs or magazines and watch food shows looking out for information on food products that people can enjoy and the availability of markets. Most of the skills possessed by creators of food products are derived from a strong educational background in science more so in fields such as food technology, nutrition, food studies and food science. They need to possess and thrive in a team work atmosphere as they will have to work with other employees seeking clarification among other things. They also need to possess good communication skills. The communication skills are useful when engaging with consumers to identify the kind of products they want or when seeking advice from other Relevant stakeholders (Caporgno & Mathys, 2018).

Creativity is also key in food production. Food developers need to brainstorm ideas and after deciding the new product to be developed, they discuss on what is required while determining the market and overall feasibility in a given product concept. They need to know how to make the new product sell by making it unique. This is achieved by research and other surveys that enables them to focus on the concept and the target goals. This enables them to define the key elements of the new food products such as vitamin fortification, consistency, color and taste.

Skills also involve the employees knowing the type of ingredients that can be used in making food. They will have knowledge on what ingredients go well together and what does not mix. They can also develop suitable food formula. They can also the flavor of food developed, its shelf life and packaging. During the development of new food products, the workers will be able to assemble and analyze the prototype of the product observing and evaluating the interaction of ingredients and the balancing of technical realities of the product within the market goal. If the product does not attain the required quality, the worker should know the kind of modifications to be made to improve the product successful. After the successful development of new food products, some workers with skills on product markets are involved in determining the final details of the product formula including packaging, pricing, marketing, shipping and distribution. They should also know how they will make profit from the new product.

However, a great challenge is that often at times there is inadequate numbers of skilled individuals. This creates a large challenge as the skilled individuals are needed at every stage of development. The absence of skilled individuals makes the creation of new food products practically impossible. When hiring workers, the company needs to access their knowledge and skills (Lamine et al., 2012). Depending on the type of food being produced, there are different workers. If food being produced needs high processing then the company will need to hire individuals skilled in food nutrition, statistics, machinery, processing and production. When using machines in food products, the individuals will need to possess technical knowledge and skills.

Figure 5.3: Consultants can give useful input on new food products.

Source: https://www.fiverr.com/hellomacrame/consult-on-food-industry-such-as-food-product-development

There are cases where companies are not able to get sufficient numbers of skilled individuals. They are forced to train some of their employees to acquire the required field. Also, there are changes in the technological world that affect food production. For this reason, workers have to be trained continuously to improve their technical knowledge. Training may require the company to hire skilled individuals from other companies, seek the advice of consultants in the food industry or if there is an individual in the company with more knowledge, he or she will be involved in the training process. Training of employees may cost the company. Some companies have modified themselves to train the employees without compromising funds to be used in the development of new food products. Though training is costly, it goes a long way in preventing wastage of resources (Kluytmans, 2010).

5.3 IDEATION

The first step in developing new food products is ideation. This process involves all relevant individuals being involved in coming up with ideas. Though idea generation sounds easy, there are companies known to hit roadblocks during this process. There are various reasons why companies reach the road block. This includes the lack of actionable Intel on product potential and the consumer and in other cases, ideation is caused by bureaucracy and disjointed workflows. Ideation cannot be done in a short amount of time as ideas have to be thoroughly thought through. Large amount of time will be needed when the team is made up of new members or when an individual from another sector wants to join the food and beverage industry. As the saying goes, great ideas take time. Therefore companies need to allocate sufficient time to facilitate the generation process. By thinking of new ideas ahead of time makes it easy to decide on new food products (Chen et al., 2013). In most cases, creative individuals are involved in the idea generation process. Creative individuals are useful in coming up with ideas on the new food products and successful marketing strategies. There are cases where companies hire individuals to involve them in the idea generation process.

Figure 5.4: Ideation is the first step in the development of new food products.

Source: https://www.shopify.com/blog/product-development-process

During the idea generation process, those in the managerial position come up with new products through the creation of suggestion bins useful during brainstorming sessions. This allows individuals to dig deep and look into various opportunities while generating ides. This is used as a repository in gathering interesting opinions from ream members and other product specialists. To encourage further creativity, companies may put in place a bonus system used in rewarding tem members with the most lucrative ideas. The rewarding system takes the form of motivation for members of staff to take part in the brainstorming initiatives actively (Khan et al., 2013). Usually, those in upper management are advised to allow their underlings to participate in brainstorming sessions before the development process commences.

To make the ideation process easier, those on the process are advised to conduct the much needed research. Research is useful in gathering information. There are various ways in which information is gathered. One of them includes surveys. Surveys can be done through questionnaires or interviewing consumers. Questionnaires used in this case are structured such that the question will give results relevant to the study. If the company seeks to change and improve an existing product, knowledge on the negative attributes of the product are used as pointers on what needs to be improved or changed. There are cases where companies get ideas on new product by looking at food greatly known and those that people would like to purchase in groceries but are not available in several stores or during certain seasons. When conducting research, the PMs are required to take responsibility for research so as to discover ideas developed by other countries and identify gaps when exploiting it.

Figure 5.5: Research is very useful in providing insights on new products that could be developed.

Source: https://www.google.com/amp/s/theconversation.com/amp/food-and-drink-products-with-pro-environment-ecolabels-are-more-appealing-to-shoppers-new-research-157260

There are cases where companies will interview individuals form different ages, races and cultures while in some instances, the group of individuals involved in the interview are from a particular group depending on the type of product being developed. Questions asked in the interview may present ideas on new food products. There are cases where individuals are called to try out old products and review them shedding light on possible new products (Joseph & Turner, 2020).

Once research has been conducted, the results will present a trend that can be interpreted used in idea generation. Usually after research has been conducted and the results analyzed, the company will gather all the members for an ideation process. During this time, they will try to come up with at least two new food concepts. The company may try to work on the developed ideas but should not be tied to the ideas. While developing the new concepts, they should try to think of new ideas. The developed ideas usually gives the company a place to start. It is very important for managers and members of the company to be open to new ideas. This creates a sense of safety for employees as they will feel free to share their ideas. As each one of them shares an idea, it could lead to other ideas. Managers should encourage members involved in the ideation process after thorough preparation. If possible, they should be notified a head of time before the actual process is done. In cases where the company gets stuck in the ideation process, the alternative is to use the Mind mapping or SCAMPER methods available. All these methods are useful in improving the process. Ideas can also come from products that were initially used and popular among consumers and were abandoned later on and no longer used.

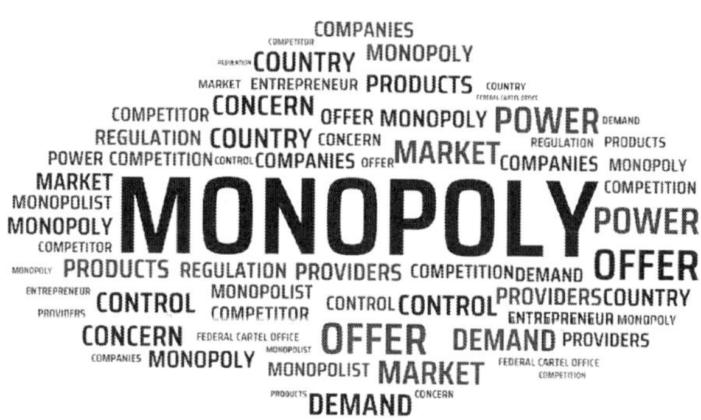

Figure 5.6: Monopoly systems present a challenge in the ideation process.

Source: https://corporatefinanceinstitute.com/resources/knowledge/economics/legal-near-monopolies/

There are various challenges faced by companies in the ideation process. Most of these challenges are faced by startup companies as already existing companies have experience with ideation. In most startup companies, managers may not have knowledge on how the ideation process is conducted. This leaves them vulnerable to making several mistakes in the process. In most cases, the startup companies may not have sufficient money to allow them hire consultants to help them in the ideation process. The lack of adequate knowledge may cause leaders in such companies to leave out employees in decision making processes (Jankovic et al., 2010). Startups and existing companies using the monopoly system may also face a challenge in the ideation process. This is because there are certain individuals involved in the ideation process. This limits the contribution of ideas. Other systems would allow employees be involved in the process as their input could yield great results. For monopoly companies, they can hire consultants to help them or choose creative employees to be involved in ideation processes.

5.4 MARKET AND PRODUCT VIABILITY

Once an idea has been generated, it may sound good on paper but present great challenges in the market place due to the lack of a financial promise. Analysis of a product's market viability is very essential before it is developed. Market viability of a product is well understood through market research. Such research is useful in providing information on the consumer preference and competition. For this reason, heads of companies are advised not to create products just because their competitors are doing the same thing. When tempted to go the same route, the companies should instead place much focus on novelties so as to attract more clients. Market viability can be achieved by conduction analysis with competitors so as to gather information from the target audience. Information on market viability can be achieved by collecting information form clients on their interests and opinions. Results from the research is used in creating a minimum viable product.

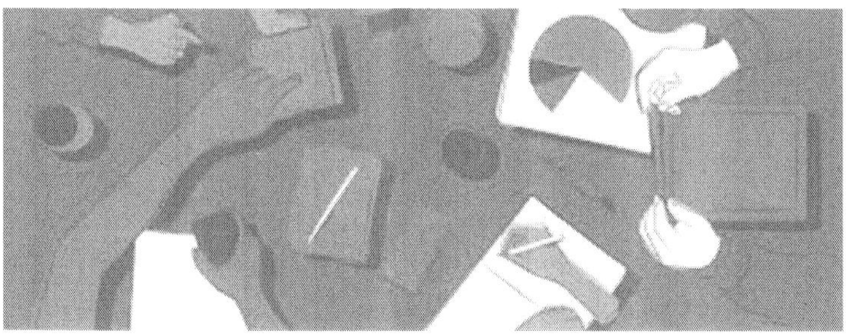

Figure 5.7: Market viability is predicted by conducting market research.

Source: https://www.bigcommerce.com/blog/evaluate-product-market-viability/

Market viability will require the company highlighting the product's core features so that consumers can identify the benefits of such products. Market viability can be predicted through the use of prototypes. The prototypes are used in getting opinions from consumers regarding the product. However, most companies have not been able to fully use prototypes in getting accurate market viability. This is because they tend to develop prototypes with only one feature contributing to the common MVP misconception. They are advised to add all the functionalities of the new product so that the consumer or investor gets a clear image or understanding of the product. By facilitating proper market viability, the company can take into consideration three main factors that influence the success of the product (Hirschi, 2012).

The factors include market size where they question whether existing market are large enough to accommodate new sellers and whether it is able to grow. It also include the targeted audience. When developing food products, there are cases where the product is meant to meet a certain group of people. Under the targeted audience, the manufacturer check whether the customer have a discretionary income or whether they can afford to purchase the products. The last factor is competition whether the company looks in into other retailers in the market, the existence of a similar product, the strength and weaknesses of the new product and the ability of the new product to survive in the existing market.

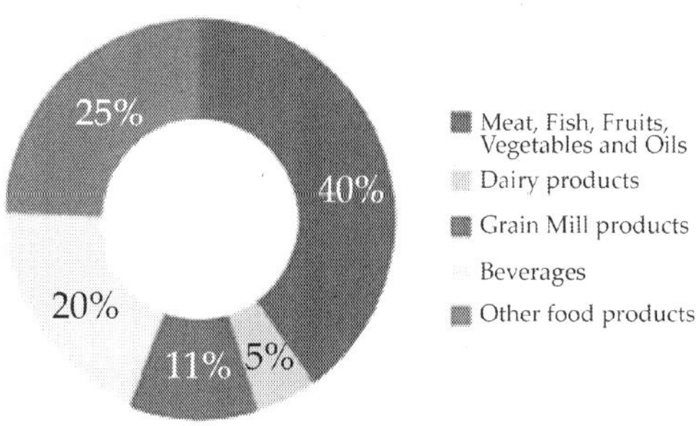

Figure 5.8: Market sizes affect the viability of new food products once released.

Source: https://www.researchgate.net/figure/Food-processing-segment-and-market-share-FY12-see-online-version-for-colours_fig2_283322011

When developing the new product, the main goal is to ensure that the product is sellable in small markets and that consumers are able to afford prices of commodities.

Market viability is often linked to product viability which the business potential of the new product is being relevant and interesting to the consumers. While developing a new product, the main goal is to ensure the product captivates consumers so that their products sell off the shelves. There are two main factors used when evaluating products. It include demand in that after the product is developed its ability to develop an interest and creating a business around it. The other factors is profit margin that cover the price point of the new product that will enable the company to generate a return on sale (Han et al., 2018). Market viability of new food products is greatly affected by food boards in various countries. In most cases, the food boards established by countries give a seal placed on all product to show that they are certified safe for consumption. For companies to get the seal, their products have to undergo a series of tests to show that they are safe for consumption. The presence of such seals in products is a major influencer of its market viability. Seals are gotten after the company develops the product and applies in the given board seeking certification. Analysis and tests are used to prove that the products are safe for consumption.

There are various reasons why companies are advised to make an analysis on the market and product viability. When developing new products or existing products, different e-commerce and potential product ideas are important. This places much emphasis on the need to evaluate to viability, profitability and customer requirement for each idea. This is very important when an individual wants to start a business in the food industry or other industries. The analysis is often done to reduce the risk of the business losing money. Under deadstock, the company evaluates the manner in which the new food product will be designed, source of raw materials and method of manufacture or processing. By making the determination, the company can be able to estimate the significant upfront investment need. Inventory management mostly applies for old products that are being reinvented allowing the company to manage it. Marketing is also essential.

Most companies both old and new place a significant amount of funds into marketing of products. Larger amounts will be spent with new products. This is because consumers have no knowledge of the new product. Lots of efforts will have to be put in place to ensure that consumers are made aware of the new product. There are cases where the amount of money spent on marketing is spent over longer durations of time. A number of startup companies do not survive past marketing as they may not have sufficient funds to facilitate the marketing of products. Analysis on market viability may highlight some of the risks linked to the production of new product (Costa & Jongen, 2006). There are various elements taken into consideration when checking product viability. The first step involves research on current markets. This involves evaluations on the depth and nature of existing markets. Research on markets is easily done through the SWOT analysis. This kind of analysis evaluates the likely strength, weaknesses, opportunities and threats of the newly developed food products. Market viability is also affected by currently market trends. Current market trends influence the type of food products being created. When the new food products are in accordance with the market trends then high levels of market viability will be observed.

5.5 PROBLEMS WITH PRODUCT ROADMAP

For the head-scratcher in PMs with no experience in extensive development experience will have a challenge in creating a product. In other cases, the PMs lacks all the proper tools needed in curating a product strategy efficiently hence the entire process is slowed down. Developing a product roadmap may present a challenge for individuals with no experience or the struggle could be in managing the product roadmap with a clear plan and vision. In most cases, companies have to grapple with roadmap prioritization so as to get the best put of the initiative. Challenges in product roadmaps can be solved by fixing the cause of the problem. In hard cases, the problem can be solved by prioritizing the OKRs or KPIs. There are cases where the MoSCoW model for products on the basis of Must-haves, should haves, could haves and will not haves. The MoSCoW model is known to save time that could be spent on brain-racking when choosing the best strategy (Granato et al., 2010). An optimized product roadmap can also be created by using various management tools such as Gantt Charts.

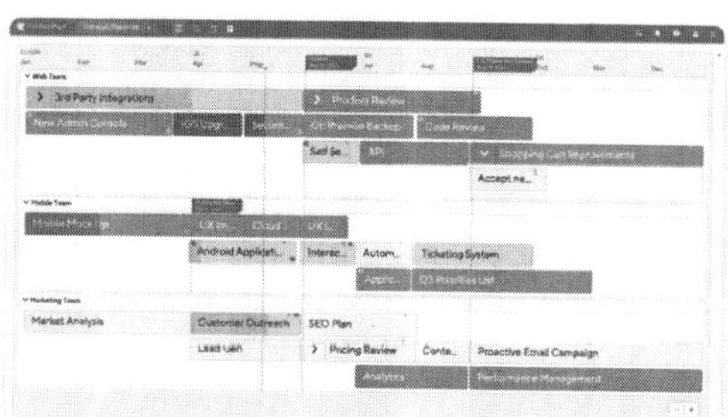

Figure 5.9: Product roadmap are useful in solving problems related to food production.

Source: https://inmoment.com/blog/5-guidelines-for-structuring-your-product-roadmap/

The MoSCoW model is known to save time that could be spent on brain-racking when choosing the best strategy. An optimized product roadmap can also be created by using various management tools such as Gantt Charts. Currently, with the use of technology it is very important to ensure that the roadmap should have a successful software development project. Such kind of a roadmap will enable the product manager to define the trajectory of a product, communicate the progress of the product to stakeholders and also visualize the company goals while justifying the changes to the budget. Product roadmaps create an avenue where not tactics and strategy combine to help members of the company in building better products (Granato et al., 2010). In as much as software development has been identified as an essential part of roadmap products, many companies have been known to struggle when implementing the product roadmaps. This attributed to the fact that road mapping is complicated in nature while the software projects. The product team has to be compatible with system being used.

Even though the companies will face difficulties in road mapping, they are far from being doomed to failure. There are some common methods used in solving issues pertaining product mapping. In most cases, when developing a product map, manufactures tend to put fixed dates and scopes on the road maps which is not right. By placing guess-time release dates and scopes parameters at various points along the roadmap is not highly effective in ensuring that the company stays on track while developing the new food product. In most cases, setting of fixed dates and scopes with regards to the release of new food products may cause the employees to sacrifice the quality of the product so as to meet the roadmap (Gaudin, 2017). The best option is to avoid the placement of such dates. This will allow sufficient time to be used in the creation of product while ensuring high quality of products is maintained. However, it is very crucial to ensure that all processes are done at a convenient time. Rather than using the dates, they can be replaced with themes. This may involve managers deciding on focus themes per quarter and work towards theme priorities while eliminating the need of deadlines. If the company is working with investors who will required constant feedback to investors, they may conduct evaluations and create looser time horizons.

Figure 5.10: Setting of deadlines may affect the quality of produced products.

Source: https://www.google.com/amp/s/www.columbusglobal.com/en-gb/blog/the-roadmap-to-new-product-development-in-the-food-industry%3fhs_amp=true

The use of deadlines can be eliminated through the introduction of project milestones and significant events. In such cases, the managers have to hope that certain tasks have been accomplished. Use of milestones have been known to help make progress very visible for the team and stakeholders. Road mapping problems can be solved by having a solution centered outlook. While developing the product roadmap, it is very easy for companies to get caught up in various elements of the roadmap. There are cases where the company reaches a roadblock in the roadmap prompting them to develop solutions (Dobrucka & Cierpiszewski, 2014). There are cases where the solutions are developed without proper understanding of the problem in its eternity. This means that the solution is not effective for the problem. For this reason the roadmap has to be suitable for the company such that it allows better understanding of the problem. In most companies, the developed a problem roadmap. This kind of problem roadmap places much emphasis on solving problems rather than creating solutions for the problem. Solving the problems involves research teams validating the problem and solving it.

While developing a food product, after the idea has been generated, everyone has an idea of the final product and how it is consumed. While developing the roadmap, it is important for all mangers to ensure that they listen to the requirements of all stakeholders and including their input into the roadmap. The challenge however comes when the manager listen to input from all the members as incorporating it into the roadmap. If everyone's idea is incorporated in the roadmap leads to the formation of a feature of soup. If the roadmap is unrealistic and also have detailed features then it is a recipe for disaster (Dimkpa & Bindraban, 2017). Though it will appear promising at the beginning, some features will be dropped in the course of development leaving some stakeholders and investors unhappy. For this reasons, managers are advised to resist the temptation of packing the roadmap with features despite the several requests that may have been made. When stakeholders are very hands on and want their input in the roadmap, they should consider organizing a dedicated roadmap visioning workshop. They can also gather ideas from all employees and stakeholders and later on develop a roadmap that complements the needs of the user.

In some cases, failures in roadmaps are as a result of poor communication. Communication is very essential in developing an effective roadmap. It creates an avenue where manufactures can agree on the kind roadmap to be created.

5.6 WORKFLOW MANAGEMENT

While developing new food products, the synergy between the collaborating teams is known to affect the quality and speed of the final product. The product manager should ensure that all members involved in development of new food products are in the same page. This can be achieved by creating an environment that requires team work. All members should be equal opportunity in access to resources or voicing their opinion (da Silva Vaz et al., 2016). Workflow management often involves communication among every involved individual including every member of the organization. This places much emphasis on good communication skills. It enables individuals to resolve conflicts between the independent teams even in the absence of team lead.

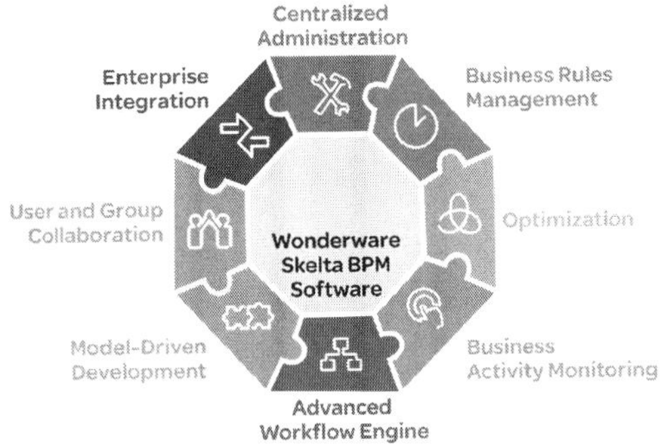

Figure 5.11: Workflow management ensures an effective production of new products.

Source: https://www.foodonline.com/doc/workflow-0001

Workflow is very essential in the food industry as any slip up could result in the production of harmful food or the new product could not be accepted by the relevant individuals. Good workflow management often at times involves the use of tools such as Trellis and Jira in organizing operations surrounding the product. This may involve the creation of a dashboard used in tracking product contributors and task completion progress. Workflow management is very essential as there are companies where workers are working remotely. These tools facilitate communication between workers while at home. In most cases the tools used are more data-intensive monitoring systems when tracking staff performance. Distribution of information among all the workers is very important (Yang et al., 2017). In some cases, information is given according to the relevant clearance. Some companies have adopted the use of a fast-paced feedback loop. There are numerous components of work flow management systems categorized in to the routing system. This kind of system is considered the basic function of workflow management. It conduces to the routing of flow of information or document flow. It also includes the transmission of information from one work item to the next. This kind of system may not be useful in exceptional circumstances.

There is the distribution system which is a function of expansion. This kind of system is useful in detecting exception circumstances and transmits information to designated work positions creating a balanced and continuation of workload within the work flow. The coordination system controls all concurrent activities preventing resource conflict or priority conflicts. The assistant system may involve methods of artificial intelligence. The workflow systems can also be categorized according to the functionalities. The categories include integration-centric workflow system, human task-centric workflow systems and XCFG. With the use of technology in developing new products, there is the automated workflow management. This kind of automation involves the management of automated system. A good example is that initially, much of the labor was provided by human beings. This labor is currently provided by using machines that have made labor cheaper and highly productive.

In the food industry, machines have been preset to conduct various activities. They can pack products into the respective package with high levels of accuracy. Machines are used in other unique ways to make work easier for all. With the automation of good production, there are various components taken into consideration to ensure workflow. Automation is mostly used after the idea has been developed and accepted hence mass production of the product. Each component is keyed to a system that makes use of data that will help users do their jobs. The first component includes materials or input needed to complete a task of food production. This component involves the HR of the company. Input components may include a contract, new hire contact information and documents outlining the kind of software or hardware equipment needed in food production (Wyrwa & Barska, 2017).

The next component is the transformation or specific actions that when applied transforms inputs to the next stage. This may involve a group of people being assigned to a certain team to perform a specific task. The third component is the output of the transformation process. This component includes the developed products being packaged and released to be sold to shopping centers and other outlets. The output component usually signifies the end of workflow. The components of workflow can be translated to rules, objects and events that generally makeup the language of the workflow automation system. The components can also be assembled into tasks, assigned to workers and tracked to completion making ideas into real objects that can be sold. It is important to note that depending on the type of Food Company, there are different types of workflows that can be placed into two main groups. They are the sequential workflow and the parallel workflow.

Under the sequential workflow, each of the sets is dependent on the previous step being completed. For instance, in the development of new food products workers in factories will have to wait until managers generate an idea and the idea gets approved for generation. It makes of rule-based workflows that makes use of a conditional logic of one action is followed by the next. The rule-based workflow is an example of the sequential workflow. The second category is that of parallel workflows (Wenzl et al., 2006). This is when multiple tasks are performed concurrently. The food production company could generate two ideas. While the ideas are being actualized, they could be generating new ideas. The parallel workflow is at times referred to as the state workflows. These workflows may at times be dependent on each other.

5.7 PRODUCT ENGINEERING ISSUES

Among the much needed skills in the development of new food products is engineering skills. In the development process, there are several engineering dependencies and problems in engineering could hinder the development process and the progress in product initiatives. There are cases where food companies want to create apps but the design team misses the deadline. The development team therefore has to wait for the prototype to be developed. These problems have to be solved on time. They are mostly solved through the establishment of a clear review cycle featuring members of inter-departmental team (Vidović et al., 2022). The problems can also be solved through the identification of non-dependent initiatives and working on them till the gridlock is resolved. This tactic enables product managers to rescue time lines ensuring that product development on track.

Figure 5.12: Product engineering affects development of new food products.

Source: https://www.pinterest.com/pin/738097826431119853/

This just goes to show that product engineering is very important. In the food industry, food engineering is a systematic, academic and expert field that makes use of science engineering and mathematics principles to food business and operation. Under food production, product engineering includes processing, handling, production, storage, conservation, distribution, control and packaging of the newly developed food product in food industries. Some of the border disciplines in product engineering include civil, chemical, industrial, mechanical, electrical and agricultural engineering. Product engineering is considered a multidisciplinary and narrow fields known to foster technological growth in food industries (Vigani et al., 2015).

There are several applications of product engineering in the food industry. It includes packaging where the newly developed food products are packaged to be sold in different markets. Food packaging innovations have often used when expanding the timeframe of realistic usability of items, settle food and ensuring that the product is perfect and speaks to the customer. Good can be packaged by packing food in jars or other methods. As developing and creating new food product makes a lot of waste, several organization have switched to using an eco-accommodating building to protect climate, therefore, drawing the attention of ecologically cognizant buyers. A few earth-arranging bundling involved the incorporation of plastics produced using potato or corn, bio-compostable paper and plastic items known to deteriorate or the reusing of content in the food industry companies. Though climate is affected by the change to eco-accommodating kind of packaging, there are numerous companies that have found this an advantage as it has led to the decreased overabundance bundling material which demonstrates the company's care on the climate.

Production engineering also considers the kind of energy used in food processing. While producing new food products, a requirement is energy effectiveness and waste warm recuperation so as to manage food preparation. This may involve the substitution of ordinary energy-serious food measures with innovations such as non-warming cycles and thermodynamic cycles. Such innovations allow the decrease in energy utilization, diminishing cost of creation and improving food creation management. It also involves the freezing and refrigeration where the principle goal is to safeguard food materials' wellbeing and quality utilizing technological food development. The freezing and refrigeration contributes to the safeguarding of short-lived nourishment and protect food quality factors such as the visual appearance, surface, taste, flavor and healthful substance. Freezing of food will hinder the development of microbes considered harmful to consumers (Veeck, 2010).

Production engineering also involves control of heat transfer during food processing. Heat management in food products in very important as it affects its. Once food products have been developed, it may be noted that some will thrive in hot temperature while others can't. Heat is very imperative in protecting food by attaining sterility, healthful and tangible food characteristics. There are different warmth move strategies that incorporate radiation, convention and acceptance. These strategies have been used to make properties of food varieties when freezing, preparing or profound broiling items and apply resistant warming or infrared radiation to new food products. All of these tools allow food designers to enhance the creation and changing food items.

Production engineering also allows the management of food safety systems. A food safety management system is defined as a methodological way of dealing with the control of sanitation danger inside a business guaranteeing that food items are safe for consumption. This is very important for new food products as it will affect is viability among consumers. There are countries that have set up a food safety management system law prerequisite that obliges all food creation organizations in the utilization and maintenance of food standards (Van Huis, 2020). In the food industry, both new and old products are usually subjected to an administration framework that covers food handling through the examination and control of the nature and substance of food while considering the risks of the food production network.

In the food industry, there are several challenges in production engineering. It includes manageability where food designs affect climates in various regions. This is through the emanation of enormous amounts of waste and contamination of air and water all of which should be tended to by food engineers on the advancement of the creation and preparation of goods. For this reason, researchers and designers are conducting all manners of tests to improve cycles reducing contamination (Tarasuk, 2008). For this reason the engineers may rethink flow practices and advances to zero in on expanding proficiency and profitability while reducing the utilization of both water and energy while diminishing amounts of wastes created.

Production engineering also affects development. There are various reasons for the creation of new food products. Among the reasons is to increase the amount of food available for consumption. With the ever growing population, continuous supply of food substances. New food products could fill in the gaps for the needed food products. For this reason, the engineers usually strive to address issues of water and land shortage giving enough development band food to undernourished individuals. Both water and land should be properly used to meet demands. The challenge is poor management of resources at various levels of production. Theories of food products to bc produces are often no practical. Such problems can be solved by picking practical approaches to food production.

Another challenge of production engineering is ensuring that the innovative food products are safe for human consumption to ensure the well-being of individuals. Production engineering usually work with experts from different spaces such as natural chemistry, commercialization and science. All food products should ensure the well-being of human beings (Sousa et al., 2008).

5.8 PRICING POLICY

It takes a lot of effort to turn an idea into an actual product. Another grave task is putting a price on the product more so if developing an agile methodology is considered. When the product is priced highly then few or no individuals will purchase it. On the other hand, charging the product at extremely low prices may lead to the company incurring losses. The company should therefore aim at finding a sweet balance in pricing of products. In most cases, the product manager is involved in identifying the maximum amount for a product that could reach the population while recording high sales (Silva et al., 2020). There are several considerations when putting a price on a product. It included hiring expenses, marketing expenses and product realization cost. The product manager so needs to consider the volatility of the initial prices. They can establish a cost estimate through analysis conducted by competitors' pricing policies.

Figure 5.13: Pricing of products affects how consumers perceive it.

Source: https://www.pricebeam.com/food-products-pricing

Pricing policies affect the production of new food products as it affects their ability to purchase the product. Pricing polices affect the purchaser and the seller as. Several factors come into play when determining the price of products. The combination includes favorable market trends, consumer liking, and product quality and product differentiation along with correct pricing. All of these factors are used in generating sales leading to the success of the product.

Pricing is different depending of the level of sale. When setting prices, producers need to ensure that they profit from selling products to wholesalers and retailers. The wholesalers and retailers also consider a number of factors when setting the appropriate price of the product. They do this be identifying their competitors and identifying the difference between products and services offered by competitors (Siegrist, 2008). However, savvy consumers comparison shops, pricing of services or products lower than prices offered by competitors may not necessarily give the seller market advantage. The manner in which consumers react to the product mostly influences pricing. Once the product has been determined as something that consumers need and want, they can considered other factors. Competitors may opt to sell their products at a lower price to outdo or match the price. A good option for producers is to convince consumers that the new food product is better than others and is of good quality.

Another essential factor taken to consideration when determining the price of products is the general expense of production. This may involve managers calculating the total amount of money spent in production then adding the amount of money they would like to make as profit. The markup is usually the difference between the amount of money charged and the price paid. The calculations include price cost of overhead and other expenses of selling. The total amount may vary depending on the food products being offered and often at times included commissions and salaries for merchant account fees, personnel and website hosting if the product is being sold online. Consumer demand also affects prices of food products. Lower prices tend to create a high demand for food products. If prices are way too low, the company will create more demands than what the company can handle.

Even though low pricing attracts large numbers of customers, it has several effects. Low prices is known to affect the volume of sales. It can either be up or down. It has been established that there are cases where retailers deliberately price certain products more so new products at low prices so as to get the attention of consumers whom they hope to sell other more expensive products. The challenge however is that when prices are low, there are consumers who fear that the quality of the product is poor. There are consumers who look at the value of the product and not just the lowest price. For this who want good quality products, they are willing to pay the price for what it's worth (Sanguansri & Augustin, 2006). This means paying large sums of money for the product. There are cases where setting of low prices can convey a message to consumers that their product is not as good as other products on the market. Setting of low prices may not earn the company great profits, the more the product sells the more the profit made. This will help improve the popularity of the product and gradually the prices can be modified accordingly. However, it is not a guarantee that low pricing will generate profit that will cover cost. This is the case when sale volume is low.

Research on the area where the products are being sold may influence the price of the commodities. For instance, there are regions where high class members of the society live. In such areas, product of low prices may not sell properly as they will be perceived to be of low quality. In regions where members of the low class live, low price products attract a lot of customers. For this reason, most of the products are made with a targeted audience. Depending on the targeted audience, prices of the commodities will change.

For newly developed products, the fact that these products are unique it acceptance varies accordingly. This means that it is not a guarantee that the product will generate high profits or customer interest. If the new food products are highly unique them it may not do well in the market if it is not liked by consumers. High pricing of the product does not matter whether the product is unique or whether consumers like it as they won't buy it. The right pricing has to be used. Some critical elements include product differentiation more so when justifying high prices if commodities. When pricing is not properly done, the consumer could price themselves out of the market. For this reason producers strive to offer high quality product whose prices could capture a significant amount of profit.

5.9 PACE OF INNOVATIONS

Development of new food products is greatly dependent on the rate or pace of innovation. This is because the exponential pace of innovation is the drives the speed of change in the market place. Companies that have mastered the art of fast innovation have been able to make use of the markets accordingly. Some companies however, have faced the problem of pace of innovation as they have slow-paced review and feedback cycles. In the food industry like other industries, product innovation is defined as the development and market introduction of new, redesigned or substantially improved goods or services (Rudolph, 2007). In the case of the food industry it involves the production of new goods. Product innovation is not only about the development of something new and original, it also includes taking something that is already existent and making it much better. This means that innovation can include taking an existing product, working on its short comings to develop something new which is an improved version of the original product.

Figure 5.14: Speed of innovation should meet to that of food markets.

Source: https://www.boardofinnovation.com/blog/40-food-innovations-that-excite-us/

In as much as product innovation is very important, it presents a challenge in various industries including the food industry. The process of product innovation is filled with detours and roadblocks. Some companies have their own crash-and- burn stories. Companies that don't know how to bounce back from failure may not be successful in production. This is because they will be haunted with the fear. Innovation has two sides to it. There is the development side which focuses more on coming up with new products. The other side is the demand side which involves the proliferation of more consumer choices than before. It is usually spurred by a multitude of digital channels brought about by changes in customer expectations. This fuels the demand for greater personalization, marketing ability, advertisement and development of products that suit consumer preference.

Changes in consumer preference on food products may cause shifts and pressures of companies that may throw companies off their game. This is as a result of uncertainty and reactivity. In such a case, the company is expected to develop a deep sense of understanding and responsiveness needed in the situation. Studies on the market show that there is a true customer need for which there is a solution for the problem. Identification of the consumer need and developing the best solution for the problem can be substantial boon for businesses in various ways. The difference in success rates is influenced by the manner in which the companies innovate, their mindset, processes and methodologies. All of these factors affect the acceleration time to profitability of the new or revamped product and the long term adoption of the product in the market (Abbas et al., 2010).

Innovation of new products has caused shifts in domination of food markets whereby brands known to dominate the market are replaced with other companies as the new products places the companies on the map. Innovation is said to be a common languages spoken by both manufacturers and consumers. Innovation is very key in the food industry as it delivers significant benefits. It is considered a critical skill in achieving success in a business. It is useful in solving problems, generating profit, increasing market shares and edging out competitors. Some of the practical benefits if innovations include improved productivity, reduced costs, increased competitiveness, improved brand recognition and value, new partnerships or relationship and increased turnover and improved profitability.

Innovation in the food industry has enabled food companies to problem-sole and provide creative insights enabling producers look at things from a different perspective regardless of the task which includes development of a new product, refreshing the strategies or finding a way to produce new and unique products to enable them stay ahead of the competition. With various companies embracing innovation in food, companies have to work constantly to develop new products. However, with innovation producers should avoid making wholesale changes mid-way into the development initiative.

Problems with innovation can be dealt with by employing employees with high levels of creativity. They can easily come up with ideas on new products and new method of marketing. Research is also useful in boosting innovation. This is because research is useful in highlighting gaps in already available products. This enables manufactures to identify the much needed products or some of the improvements that can be made on already existing products.

5.10 COSTLY MISTAKES

Among the challenges faced in production of new food products is when manufacturers make costly mistakes. In most cases, the mistakes are made with regards to the technology used, systems used, food combinations or formulation of ingredients in the products. If the manufacturer has an issue with a fermented products, they may have to get the right viscosity during the storage-life needed. This may include management of the microbial fermentation to attain the required levels of acidity, adding the right amounts of the right types of thickening agents or controlling temperatures needed in gelling. When a problem occurs, right measures should be put in place to solve it (Roy et al., 2009). This often involves an investigation being conducted to understand the product. For instance, with the fermenting products, the problem can be solved by understanding the product's microbiology, formulation of ingredients, interaction of ingredients and control of processes. This ensures that the desired texture of the product is reliable.

Costly mistakes can be avoided by enlisting the right expertise that ensures a smooth transition from concept to product. This is very important as it saves the company money through reduced waste production and use of energy. It helps the company deal with slender profit margins which is a norm among manufacturers in the food and beverage industry. The food and drink industry is very unpredictable as problems can arise from time to time. Quick identification of the problem and dealing with is effective and as soon as possible jn useful in reducing downtime and saving money.

5.11 TIME-TO-MARKET

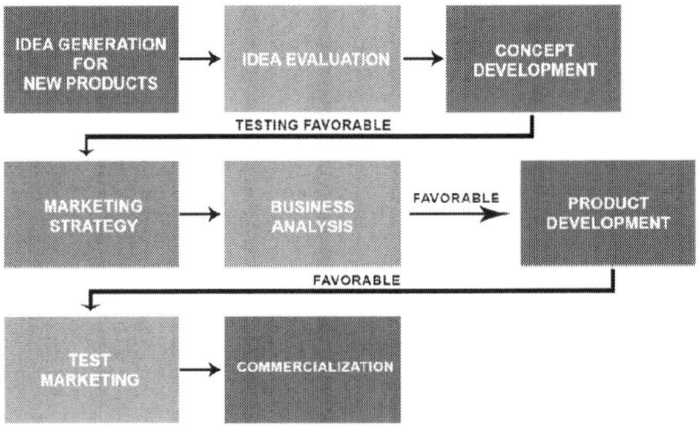

Figure 5.15: Products should be timely produced to attain the required goals.

Source: https://www.intechopen.com/chapters/59751

Once a product has been developed, its market positioning is determined by the company's ability to meet deadlines and launch timelines. Project managers should be able to fine-tune operations. This is useful in ensuring that all activities goes according to the established timeline. The issues mentioned above may affect the overall time-to-market, there are roadblocks that may cause the development process to reduce significantly. For some of these issues, the project manager cannot deal with them while others are caused by gross incompetence within the company. Several factors affect the time to market. They are discussed as follows:

There are unforeseen circumstances that included the COVID-19 pandemic. Since its first occurrence, it has affected several businesses across the world. The pandemic has caused companies across the world on overhaul the entire workforce management while addressing the changes. Various businesses including hotels have had to abandon products so as to accommodate the financial impacts of the pandemic (Roohinejad et al., 2017). Some companies were not able to acquire investors or maintain them as large amounts of funds were directed towards dealing with the pandemic. Also setting low prices of commodities resulted in companies incurring losses as most of the ingredients were expensive. Setting of high prices was also a problem as most people are not able to afford high prices of commodities.

The legacy system also present challenges. Although they are very cheap to maintain, they have high levels of unpredictability which makes them a liability in the product development cycle. Malfunctions in the legacy software system can caused the company to move backwards by a few months or longer depending on the effects. Hiring delays are also a problem. Most companies have the challenges of poor recruitment which slows down the product's launch timelines significantly. For instance, when onboarding new app or product designers work on a prototype can slow down the design process. Several companies have not achieved the teleworking levels and therefore struggle to train new hires and meet deadlines. To deal with this issue, the company should always have a contingency plan to deal with issues. This kind of strategy helps the company to survive massive scares while meeting their quarterly targets.

Hiring new employees also involves the product promotion may be noted. Here the product manager will work in collaboration with marketing teams to think of promo options. There are cases where single strategies are employed such as the use of blogs and SEO to favor combined efforts. Companies should ensure that the quality of food is maintained or upheld. Companies should not sacrifice their ultimate quality to decrease the time to market (Reis et al., 2014). The company's reputation can be damaged by the introduction of a poor quality product. This will cause consumers to estrange from the company. Companies are also advised to set objectives for each development stage and ensure that every member is able to visualize the goals in real time.

Companies have also faced the challenge of compliance issue. It includes the influx of new products into the marketplace, establishment of regulatory bodies that have become stricter on compliance and standards of quality. This means the banning of companies when they take part in illegal activity. The banning of such companies highlights the strictness of the ever-changing food industry. Though there are standards and regulations to be followed, the company can avoid the compliance issued by following the updates guidelines of the newly developed product before they enter another development stage. Project managers should also encourage teams to download industry-regulated templates for every product initiative.

Under post-launch improvement, the product's life cycle does not end after the product has been launched (Pereira & Vicente, 2010). The project manager has to plan updates while ensuring that the consumer need are met. Companies can organize outreach campaigns. These campaigns are useful in expanding the company's audience to global audience. Bulk email services can be used to inform users about new products features and sakes promo. They can also create user manual for the newly developed products.

Newly developed models need certain modifications to better serve the food industry. Some of the key areas taken into cony include high emphasis on meeting consumer demands and sensory attributes. This include contemporary consumer demands being placed in a premium on minimal preparation time and seasonal colors and palatable textures. Seasonality of ingredients is also a great consideration as consumers value seasonal foods attributed to their perception of quality and local sourcing. When producers overly rely on in-season ingredients can be very risky as their availability cannot be ascertained throughout the year. The distribution and scouring of local ingredients where there are potential benefits if supporting local networks rather than relying international supply chains. Some of the environmental benefits of controlled production include reduced emission.

Development of new food products is very important as it helps deal with various issues in the society. Some of the issues include food insecurity. The new food products fill in the gaps of missing products enabling consumers get access to various products. Production of new products is also useful in controlling food prices. As there are various options for consumers to choose from, prices of new commodities are usually low so as to get the attention of users. New food products also provide a variety of options of users such that consumers are able to benefits from various nutrients in new products. For this reason the challenges discussed above should be dealt with as the right stages (Arzani & Ashraf, 2017). Product managers should also ensure that they are in compliance with government requirements and guidelines with regards to new food products. Continuous training is advised to ensure all employees are well equipped with required skills and knowledge to put their company on the map. They should also set attainable goals with regards to production. The products should reach food markets at the required time for large sales volumes.

CHAPTER 6

FAILURES AND SUCCESSES IN THE DEVELOPMENT OF NEW FOOD PRODUCTS

CONTENTS

6.1 INTRODUCTION

Product development has been a key activity in the food business for more than four decades, but it has only recently emerged as a critical business sector and sophisticated technology. For a long time, it was essentially a craft, only tangentially tied to the company's research and engineering departments. The rising supermarkets' demands for a continuously changing, vast range of items and continual price reductions put a lot of pressure on product development (Peters et al., 2016). So, there was a push for product differentiation, including tiny adjustments that were enough to identify items on the shelf, as well as cost savings.

Figure 6.1: Food and beverage industry innovation has always been fiercely competitive, but it has always been accepted and promoted since it is critical to the sector's survival and growth. Companies like this challenge since the benefits of introducing a new product to the market are significant - both financially and in terms of the company's standing in the global market.

Source: https://www.integratedfoodprojects.com/news/top-10-failed-food-products/

When it comes to the food industry's overall successes and failures over the years, there has been socioeconomic success in providing enough cheap food in developed countries, but failure in developing such a bad reputation that the food industry became heavily regulated; industrially, there has been great achievement in developing massive multinational companies, but failure in constantly reducing food product margins. Can narrowly focused company plans, a lack of innovative strategies and organization, or a lack of expertise be linked to failures?

There are currently significant social and technical demands on the entire food system to change quickly, such as those resulting from the expansion of information technology in more affluent nations and the growing economic power of certain developing countries. Is the food sector up to the task? Are there enough expertise and people in the food industry? How will it react? The goal of this is to examine the causes of past product success and failure, as well as to identify significant challenges for future product development success (Pereira & Vicente, 2010).

What exactly are foodstuffs? What are some examples of new food products? Even while everyone believes that food is stuff that is eventually ingested by people to meet physiological and psychological demands, the food firm and the customer may have completely different definitions of the food product on the market. To create a comprehensive corporate product, the firm develops a fundamental functional product to which it adds packaging, aesthetics, brand, pricing, and promotion. The customer characterizes the product as a set of advantages, tying the product's tangible and intangible characteristics to their needs, wants, and actions.

The description of a fundamental food product, such as flour, might be straightforward and pragmatic, but it can be complicated and emotive for items such as a restaurant meal. A new product, according to the firm, has some differences in fundamental functions and aesthetic presentation; nevertheless, customers compare it to the "old" product and competing goods, and if they see a difference, it is a new product to them. The goal of product development is to bring these two viewpoints together.

Canned foods, frozen foods, dried foods, chilled foods, and ambient foods were once often categorized according to their preservation technique. Milk products, for example, are classified as "fresh," "ultra-heat treated," "canned," and "dried," while fruits are classified as "fresh," "canned," "dried," and "frozen." The fundamental rationale for this grouping was that the preservation technique was dominant in processing, distribution, and retailing; changing the preservation method required a significant investment of time and resources (PF Guiné et al., 2016).

Chilling, canning, and freezing, are all thermal techniques that use temperature and time to regulate food quality. Dry and ambient foods are preserved through non-thermal mechanisms such as regulating water activity, atmospheric gases, and packaging. There has been a surge in interest in non-thermal food preservation, such as irradiation and high pressures, in recent years.

Figure 6.2: Hundreds of thousands work in the food industry around the country. Food processing plants and factories may be found all across the country. Modern food production enables not only the manufacture of maximal goods from the smallest amount of raw materials, but also the preservation of products for longer periods of time, avoiding damage and rejections, which is no less significant.

Source: https://greaterthamesmarshes.com/food-production

Both procedures arose in an attempt to minimize food quality degradation during processing, but each has its own set of challenges. Products are further divided into categories based on processing methods such as baking, extrusion, and fermentation, as well as food forms such as liquids, emulsions, and powders. This is beneficial since it recognizes the company's core technologies and expertise of them. If the company's biggest expertise is in emulsions, then product categories include ice creams, margarines, salad dressing, cooking oils, and sauces, with new goods produced from fundamental emulsion knowledge (O'Shea et al., 2012).

Bread, rolls, and cakes; crackers and biscuits; confectionary; sauces and pickles are other common categories. A new process technology might kick off a product family, or perhaps numerous product families. Extrusion technology, for example, was used to create a variety of innovative food items, including flavored puffed nibbles and muesli bars. Product and processing knowledge is vital in product development since it may lead to substantial advances, such as "first-to-market" items.

6.2 SUCCESSES

Chemical engineers have been conceiving, designing, testing, and scaling up novel food-processing systems for over a century. We can now anticipate exotic vegetables and fruits, dairy, seafood, and fresh meats, to always be accessible on our local supermarket aisles as a result of their efforts. To fulfill our nutritional and gastronomic demands, we've grown to rely on a plethora of safe, easy processed meals.

6.2.1 Urbanization, Population Growth, and Ageing

While global population growth is decreasing, certain areas' populations will continue to rise long beyond 2050, maybe into the next century. Cities presently have a higher population density than rural regions, and this disparity is expected to widen as the population expands. Food systems have been greatly impacted by urbanization, which has coincided with a shift in eating patterns. The world's population is getting older as a whole (Balasubramanian & Panigrahi, 2011). Aging is already speeding up in low-income nations, where the process tends to begin sooner and is more prominent in rural regions. The agricultural labor force and the socio-economic fabric of rural communities will be significantly impacted by urbanization and ageing.

FAO has traditionally included population dynamics, which includes variety in age group structures, regional trends, and location, as a primary driver of changes in demand for food and agricultural products when making predictions (rural and urban). The UN Population Division has calculated population increase in three alternative scenarios, referred to as low, medium, and high versions.

Annual rates of growth in the world as a whole have been dropping for almost five decades. Global growth rates peaked in the late 1960s at 2%

annually, with total fertility rates (TFR) reaching 4.5.2. Annual worldwide population growth rates declined to 1.2 percent in 2015, while TFRs fell to 2.5. Despite reducing global rates of growth of the population, absolute yearly increments continued to rise until recently, when they began to notably drop. The absolute yearly increases are at little about 80 million people.

Figure 6.3: Food processing enables us to consume a wider range of foods than our forefathers. We can carry foods from all over the world to our homes because we can securely preserve and package them. We are not limited to what is produced locally or by season, dramatically increasing food accessibility and availability for the vast majority of people who live in cities. With more options, we may eat a more diversified diet that is more likely to include all of the nutrients we need for optimum health.

Source: https://www.futurelearn.com/info/courses/how-is-my-food-made/0/steps/63298

Global trends obscure significant disparities across and within regions, as well as between high-, middle-, and low-income nations. While high-income nations will achieve their optimum size of the population by

2040, low- and middle-income nations will have only modest growth in the medium and long periods. Within low-income nations, there are also significant disparities in population growth rates. Between 2050 and 2060, Asia, the world's most populated continent, will reach its population peak.

Should these population estimates come true, the resulting increases might threaten these countries' overall growth chances. Because agriculture is such an important source of employment and revenue in all of these nations, it would also jeopardize efforts to improve nutrition and food security. This is especially true for nations like Niger and Somalia, which are heavily reliant on agriculture and have limited water and land resources. If current trends continue, these countries might face a neo-Malthusian future if they rely only on local production for their food supply (Nicklaus, 2011).

The population structure is changing as a result of rapid population growth, with younger generations accounting for a larger share of the general population. Between 2015 and 2050, the number of persons aged 15 to 24 years old in low- and middle-income nations is predicted to climb from around 1 billion to 1.2 billion. The majority of these young people are predicted to reside in South Asia and Sub-Saharan Africa, primarily in rural regions where employment will be scarce.

For decades, the majority of the world's population lived in rural areas. More over 60 percent of the population resided in rural regions 35 years ago. Since then, the urban-rural balance has shifted dramatically, with cities now accounting for slightly more than half of the world population (54 percent). In 35 years, in 2050, more than two-thirds of the world's population may be living in cities. Agriculture has changed as a result of technological advancements and the use of labor-saving technology, which has aided in the acceleration of urbanization. At the same time, reforms introduced by urbanization have had, and will continue to have, an impact on agriculture, food, and nutrition.

Food consumption trends are influenced by urbanization. As part of a broader nutritional change, more urban wealth tends to boost demand for packaged foods, and also animal-source food, fruits, and vegetables. Higher urban wages also tend to raise the opportunity costs of food preparation, favoring food items with a high labor component, such as fast food, store-bought comfort meals, and foods produced and sold by street vendors. Dietary nutrient composition is shifting as a result of these changes. Diets are often growing more salty, fattier, and sugary, as well as more energy-dense (Nedovic et al., 2011).

This shift in consumption habits has also resulted in a shift in employment within the food system, with fewer people working in agriculture and more in transportation, wholesale, retail, vending, and food.

Figure 6.4: When compared to a vegetable-based diet, our greater intake of animal products necessitates the use of more crops to provide the same amounts of calories and protein. On area that is now utilized for livestock, four times more plant-based food may be produced.

Source: https://pursuit.unimelb.edu.au/articles/we-can-have-urbanisation-and-enough-to-eat

Rural regions experience earlier onset and quicker progression of aging than national statistics would suggest. Rural ageing has significant ramifications for the rural labor force, agricultural output patterns, land tenure, social structure within rural communities, and overall socioeconomic growth. Older farmers are more affected by environmental deterioration, climate change, and restricted agricultural technology than their younger, healthier, and better-educated peers. Discrimination against elderly rural people in obtaining loans, training, and other income-generating opportunities may exacerbate the challenges encountered by older farmers (Monteiro et al., 2010).

Agricultural advances, such as the spread of new agricultural technology and the adoption of improved seeds and implements, frequently bypass elderly farmers, who lack the financial means, skills (e.g., literacy), and energy to engage in adopting new techniques. Gender divisions in agricultural production disadvantage older women, limiting their access to credit and training, as well as their ability to participate in market exchanges.

Adapting farming technology and agricultural policies to the skills and requirements of older farmers in nations where the agricultural labor force is aging might assist to keep older people involved in productive activities. The provision of social services in places experiencing 'compressed ageing' may need the adaption of social support networks to meet the altered age structure.

6.2.2 Investment opportunities and Food Prices

Between 1990 and 2014, the global economy increased at a rate of 2.6 percent per year, nearly doubling in size. Low- and middle-income nations, whose gross domestic product (GDP) increased by 5.1 percent annually, were the major drivers of global economic expansion during that time. China's GDP increased at double that rate, at more than 10% each year, and the country accounted for 9% of world GDP in 2014, compared to only 2% in 1990. The typical global citizen's income is currently almost 1.4 times larger than it was in 1990 (Morgan & Sonnino, 2010). However, there are significant geographical variances. Between 1990 and 2014, per capita income in rising East Asia and the Pacific climbed by 7.4 per cent each year. In Sub-Saharan Africa, on the other hand, average annual income growth was a meager 1.1 percent, reflecting radically divergent development patterns between low- and middle-income countries.

Figure 6.5: In September 2021, global food prices increased by about 33% over the same time the previous year. According to the United Nations Food and Agriculture Organization's (FAO) monthly Food Price Index, worldwide prices have climbed by more than 3% since July, hitting levels not seen since 2011.

Source: https://theconversation.com/why-global-food-prices-are-higher-today-than-for-most-of-modern-history-168210

Over the last 25 years, global rates of economy-wide investments (or gross fixed capital creation) have climbed. Between 1991 and 2014, investment growth in high-income nations averaged less than 2% each year. As a result of the global financial crisis of 2008–09, these nations' investment levels plummeted by over 15%. By 2014, levels of investments had returned to roughly US$9 trillion, which had been around pre-crisis levels (at constant prices of 2005).

Over the last quarter-century, gross investment in The Pacific and East Asia has risen rapidly. It surged from US$300 billion in 1990 to US$2.8 trillion in 2014, owing mostly to China, where capital increased at a 15 percent annual pace. China contributed for more than 85% of investment in the area in 2014. Other middle- and low-income nations' investment growth has been substantially slower, averaging little more than 4% each year. While levels of investments have risen, investment rates in high-income nations have stayed practically unchanged, at roughly 22% of GDP. Between 1990 and 2015, China's economy-wide investment increased from 28% to about 45 percent of GDP; rates of investment in other middle- and low—income nations followed a U-shaped trend. Even in Sub-Saharan Africa, where investment rates are far lower than in other low- and middle-income countries, there is increasing growth after a negative trend from 1995 to 2005 (Mitterer-Daltoé et al., 2014).

Investment trends in agriculture, on the other hand, paint a somewhat different image. In absolute terms, low- and middle-income nations now invest almost as much in agriculture as high-income countries — roughly US$190 billion in both groups. Agricultural investment grew in all nation groups between 1991 and 2014, but at differing rates. Investment climbed from roughly US$120 billion to US$180 billion in high-income nations, representing an average annual growth rate of around 2%. It increased from less than US$11 billion to US$76 billion in China, representing a 9 percent increase, while agriculture investment in the remaining low- and middle-income nations increased from US$45 billion to US$115 billion, or a 4 percent increase.

The fact that low- and middle-income nations account for the majority of global agricultural investments does not suggest that the sector is viewed as more vital than it is in terms of scale. A comparison of agricultural investment shares in total investment and agricultural value-added shares in GDP indicates significant structural disparities as well as distinct trends across groupings of nations. To begin with, agricultural investment outnumbers agricultural value added only in high-income nations. In the previous two decades, high-income nations have consistently invested more in agriculture than the sector's percentage of GDP.

The 'agricultural investment orientation ratio' has maintained constantly over 1 as a result of this. This ratio is substantially lower in middle- and low-income nations, hovering around 0.4. Second, over the last two decades, distinct regional trends have emerged. While the investment orientation ratio rises in high-income nations, East Asia and the Pacific (including China), South Asia, Europe, and Central Asia, it falls in the Middle East, Caribbean, Latin America, Sub-Saharan Africa, and North Africa (Batista et al., 2013).

In agriculture, the degree of capital intensity varies as well. Agricultural production in high-income countries is markedly more capital intensive than agriculture in low- and middle-income countries, as shown in Figure 2.9: it takes four units of capital to create one unit of value added in high-income countries, especially in comparison to around 1.5 in low- and medium-income countries. However, agricultural capital intensity is expanding across East Asia and the Pacific (including China), South Asia, Europe, and Central Asia. While this cannot be read as an indication of convergence toward the sort of agriculture practiced in high-income nations, it might imply that capital is gradually substituting other inputs and components, notably labor.

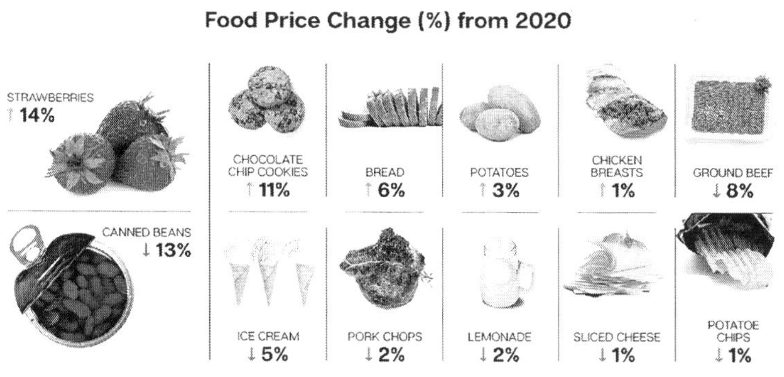

Figure 6.6: Prices for staple goods such as bread, pasta, rice, and potatoes are growing at a quicker rate than other items due to increased wheat/grain prices this year and probably next year. Meat costs, particularly pig and poultry, are also higher than in 2019. Seafood costs haven't risen as quickly as red meat or chicken prices this year.

Source: https://foodinstitute.com/focus/food-industry-financials-dont-be-surprised-to-spend-an-extra-6k-on-food-this-year/

Economic cycles in the world economy largely explain trade trends. Trade policies and agreements play a role as well, although their influence is harder to assess. The failure of intergovernmental trade negotiations under the auspices of the World Trade Organization (WTO), particularly the inability to complete the Doha Development Agreement and a partial reversion to protectionist policies following the global financial crisis, may have exacerbated the global trade slowdown.

The Trans-Pacific Partnership (TPP), the Regional Comprehensive Economic Partnership (RCEP), and the Transatlantic Trade and Investment Partnership (TTIP) are three big regional trade agreements (RTAs) that have recently been signed or are currently being negotiated (TTIP). Agriculture is involved in all three, or at least has an impact on it. They want to liberalize agricultural commerce even further, change food safety restrictions, improve plant and animal health, and harmonize food product standards (Matser et al., 2004). These RTAs are also anticipated to address the legal rights and duties involved with the use of particular foods and wines' names in international trade, as well as the scope of plant patent protection. Additional laws on the use of agricultural export subsidies, as well as the situations under which export commodities' limitations might be imposed, are also on the table for discussion.

The nations negotiating these accords have varying levels of border security for agricultural imports, ranging from near none for Australia and New Zealand too high for agricultural goods imported by Canada and Japan. Except for the United States of America and European Union, several of the TPP and RCEP nations have already signed bilateral RTAs that have abolished or are in the process of eliminating tariffs on numerous commodity and food imports.

Some low-income nations are concerned that the removal of agricultural tariffs and quotas, which is now being discussed among RTA members, will undermine the tariff benefits that have given their food exports a competitive advantage in those countries. The amount to which agricultural products from third nations are used as inputs by each trade bloc's food processing industries might be influenced by the rules of origin developed in each megaregion. Another concern is that strengthened TPP and TTIP regulatory disciplines and processes will legitimize how regulations on Sanitary and Phytosanitary Measures and Technical Barriers to Trade are implemented, paving the way for more stringent standards that third-country agricultural exporters will find more difficult and expensive to meet.

Following the peak in 2008 and 2011, the FAO's actual food price index has dropped to levels last seen in the early 1980s, but it remains significantly above 1990s and early 2000s lows. The latest joint research from the FAO and the Organization for Economic Co-operation and Development (OECD) paints a mixed picture of medium-term trends in real food commodity prices through 2025. While meat and cereal costs, with the exclusion of coarse grains, are expected to fall in real terms over the next ten years, dairy prices are expected to grow. While prices are expected to stay fundamentally higher than they were in the decade prior to the price increase in 2007–2008, such medium-term changes are "not incongruous with an extremely long-term pattern of dropping real prices." The downward linear trend of the FAO's real food price index (RFPI) reflects this declining trend in agricultural commodity prices, albeit its statistical significance becomes less obvious than in the case of particular commodities, such as real wheat prices (Martelli et al., 2020).

Figure 6.7: According to a recent FAO study, the Food Price Index reached 123.0 points in July 2021, down 1.2 percent from the previous month but still 31 percent higher than the same time in 2020. The decreases in July continue a pattern that began in June, when the Index fell for the first time after twelve straight months of gains.

Source: https://www.foodbusinessafrica.com/global-food-prices-fall-for-second-consecutive-month-fao-food-price-index/

Following the food price inflation crisis of 2007–08, food price changes around the trend, as well as heightened volatility and unpredictability, got a lot of attention. Food price volatility appears to have peaked in the 1970s, according to standard measurements of volatility during the previous 50 years. However, since 2000, price swings have been higher than those seen in prior decades, when prices were also underneath the linear trend line. Given the reduction in the FAO index between 2015 and 2016, it appears that volatility has continued to rise. These aberrations point to volatility levels similar to those seen in the 1970s.

Future food costs will be determined, among other things, by how well production can adapt tighter resource restrictions and climatic change. Climate change may imperil the ability of some parts of the world to increase agricultural production in order to satisfy rising demand, putting increased pressure on pricing. Furthermore, carbon-emission costs may need to be internalized as part of mitigation efforts. Furthermore, costs may rise in the long run if there is a need to limit GHG emissions so as to comply with international climate change accords (Lusk & Briggeman, 2009). Adopting these mitigating measures, on the other hand, would incur higher expenses (at least in the short term), putting upward pressure on output prices.

6.2.3 Look and Taste

People have always been on the lookout for creative methods to enhance the texture and flavor, and presentation of their food. Today, a whole discipline of chemical engineering is devoted to using science and technology to improve the taste and appearance of the food we consume. Chemical engineers have been collaborating with food scientists to identify and commercialize natural and artificial tastes, as well as other food additives. As a result, much more pleasurable eating experiences have been created. Food science and manufacturing now have their own section of chemical engineering. The researchers in this field's unique, combined knowledge have been aimed toward: enhancing the tastes and textures of food, as well as adding nutritional value and improving the look of dishes. Artificial and natural sweeteners and flavors have seen some of the most significant developments. Innovative applications of various starches have also been used to improve the texture and nutritional content of foods.

Sugar cane or sugar beets are used to make refined sugar, which has profoundly influenced the natural sweetener industry. Commercial-scale refined sugar manufacturing entails a number of chemical-engineering processes, including: milling shredded of raw materials and combining them with water, adding chemicals to manage the acid levels, eliminating impurities, crystallizing sugar and drying it, and treating effluent are all part of the process. High-fructose corn syrup, which is manufactured from cornstarch, has been utilized in meals and beverages in recent years (Lamine et al., 2012). Corn syrup production also necessitates a number of chemical-engineering procedures, including: The corn is milled dry, the cornstarch is reacted with enzymes, and the corn is purified using ion exchange. Food processors choose high-fructose corn syrup because it tastes sweeter than sugar that is refined and is generated as a syrup, making it easier to incorporate into a variety of meals and beverages.

Artificial sweeteners are new chemical compounds that have a sweetness level 500 to 600 times higher than regular sugar. Calorie-conscious customers and diabetics who must restrict their sugar consumption highly respect them. Chemical engineers have spent a lot of time developing the techniques that are required to synthesis these chemicals and manufacture them commercially. Saccharin was the first commercialized artificial sweetener, discovered in the late 1800s. Sweet'N Low is the brand name for this product. Aspartame, often known as Equal and the food ingredient NutraSweet, has lately acquired favor in the market. Sucralose, which is branded as Splenda and is actually manufactured from sugar, is the most recent artificial sweetener to hit the market. Chemical engineers employed a variety of process advances to develop the commercial procedures that generate these low- or no-calorie sweeteners.

Figure 6.8: Functional meals will grow more sophisticated and popular, especially when the now-legal cannabinoid CBD is introduced. Larger M&A acquisitions, as well as a slew of smaller ones, will set the course for where Big Food will go next. Packaging and manufacturing will continue to be more environmentally friendly, and businesses will get closer to meeting their sustainability goals.

Source: https://www.helpguide.org/articles/healthy-eating/high-fiber-foods.htm

As a key source of nutritious carbs, starches from numerous sources are included into processed meals. Starches are also commonly added to dishes as thickeners, to increase stability, and to give the customer a pleasing "mouth feel." Chemical engineers were instrumental in extracting the required starch from cereal grain seeds, tubers, and roots. They are also in charge of designing complicated systems that create commercial-scale volumes of easy-to-use, free-flowing powdered starches (Kluytmans, 2010).

6.2.4 Packaging

Consumers' food choices used to be confined to what could be manufactured and carried locally. This has changed because to sophisticated packaging produced by chemical engineers, which today allows us to enjoy a diverse and abundant range of meals. Chemical engineers have contributed to the abundance of fresher-tasting, longer-lasting goods we enjoy today by devising revolutionary packaging technologies. Traditional metal cans, multilayered packages that allow heat sterilization directly in the containers, and packaging that manages carbon dioxide and oxygen levels to slow ripening and prevent spoiling are just a few of the modern packaging wonders.

Foods can be carried over longer distances and held on the shelf for longer lengths of time thanks to the protection given by specialized packaging. Protecting perishable commodities such as fruits, fish, and meat from deterioration and decay has aided in the connection of distant sources of food with new markets throughout the world. Strawberries, kiwi fruit, and mangoes are now available year-round at local stores. Sterilization, vacuum packing, and multilayered packaging are some of the most prevalent packaging methods (Khan et al., 2013).

Dehydration, candying, pickling, salting, smoking, and even the use of particular spices has all been tried to delay rotting. Food spoilage is now well understood to be caused by molds, bacteria, oxidation, and microbes, and that sterilizing can help avoid this. Many improved sterilizing procedures to preserve goods and protect customers from food-borne diseases were developed with the aid of chemical engineers. High-temperature pasteurization and canning, freezing and refrigeration, and irradiation are examples of these methods.

The development of vacuum packing has been heavily influenced by chemical engineers. The meal is placed in a gas-tight bag, which is sucked out before being sealed. This procedure limits the amount of oxygen in the bag, making it impossible for microorganisms to live. Modified atmosphere packaging (MAP) and Controlled-atmosphere packaging (CAP), both of which constitute historic chemical-engineering accomplishments in the realm of food processing, are later refinements on classic vacuum packing. To prevent fruit and vegetable respiration, carbon dioxide and oxygen levels inside the food packaging environment are regulated during CAP. This method delays the ripening and rotting of fresh food while also extending its shelf life. MAP is a more sophisticated variant of CAP that delays ripening while also preventing several natural processes that cause foods to get rancid and perish.

Figure 6.9: Every new food business hopes to see their product on the shelves of Whole Foods. However, it has caused some well-known firms to go out of business. Contracts will leave you with small profit margins but high order quantities. To expand up, you may need to invest in costly equipment.

Source: https://strateege.com/specialty-food-marketing/28-lessons-learned-small-food-business-factors-success-2/

Another significant chemical-engineering innovation to food safety and convenience is the development of chemically sterilized brick packs. These multilayer containers are commonly used to keep tomato sauce, milk, juice, and a variety of other foodstuffs from deterioration without requiring refrigeration. High-quality plastic, paperboard, and metal are used to make the innovative brick-shaped container. Flexible, laminated retort pouches have lately become popular for food preservation (Barreteau et al., 2006).

Wet meals are placed in multilayered retort pouches, which are then sealed and heat treated to sanitize the contents. Because the food inside the pouch is never subjected to the atmosphere again until it is consumed, it may be kept without refrigeration.

6.2.5 Food Convenience

Prepackaged, frozen, quick-cook, dried, and microwaveable foods are just a few of the innovations that have helped to make our fast-paced modern existence more tolerable. Chemical engineers have primarily been responsible for the advancement of today's time-saving conveniences. Heat and eat in a flash; simply add water... and you are on your way, completely satisfied. The creation of a variety of quick and appetizing dishes has aided our present fast-paced lifestyles. Chemical engineers are significantly responsible for conceptualizing and implementing the numerous automated processes required to manufacture today's convenient meals.

The removal of nutrient-rich outer bran layers was one of the first nutritional compromises made in order to produce fast-cooking beans, rice, and other legumes. Chemical engineers devised ingenious food-processing procedures that enable very nutritious brown rice, wild rice, beans, and a variety of crop seeds to be cooked swiftly without destroying their outer coats.

The first objective was to create frozen meals that thaw and cook without losing their look, texture, flavor, or nutritional value. Quick-freezing techniques were devised by chemical engineers to assist thawed foods keep all of their freshness. The enzymes that cause discoloration and the formation of unpleasant tastes are suppressed by blanching vegetables followed by rapid freezing. In the 1960s, the concept of individual fast frozen processing was established. It includes flash freezing food such that when it is thawed, it retains its inherent qualities (Joseph & Turner, 2020).

6.3 CHALLENGES

The previous examination and analysis of global trends impacting food security, hunger and poverty, and the sustainability of agricultural and food systems yielded many major conclusions. To begin with, total food demand will continue to rise, despite rising depletion of natural resources and significant changes in the structural component of food and agricultural product demand. Natural resource deterioration and scarcity will continue to be exacerbated by climate change and growing competition for natural resources, posing a threat to people's livelihoods and food security. Extreme hunger, poverty, food insecurity, and malnutrition will continue to be issues, as will rises in overweight, obesity, and diet-related chronic illnesses.

Natural disasters are becoming more frequent and intense, and they, together with climate change-related extreme weather occurrences, are likely to exacerbate the worldwide demand for humanitarian aid and rural family resilience development. Transboundary plant pests and diseases, as well as other developing hazards, are continuing to cause crises in agricultural and food systems, affecting production and human health. Conflicts are occurring and may increase in many regions of the world, with far-reaching economic and social implications that extend beyond the nations involved (Bellows & Anderson, 2006).

Most low-income nations are experiencing dynamic rural development, which is projected to continue. This will have ramifications for agricultural production systems, employment, nutrition, and migration, and society will face the issue of figuring out how to engage everyone in the development process. Rapid shifts in food systems necessitate more effective international and national governance structures, as well as evidence-based and well-targeted policy responses. To improve agricultural productivity and foster innovation for sustainable farming, rural prosperity, and food security, more investment in agriculture and agricultural food systems is required, particularly higher spending on research and development.

6.3.1 Increasing Demand

Between 2012 and 2050, demand for food as well as other agricultural goods is expected to grow by 50%. Demand will shift structurally as a result of variables such as population expansion, urbanization, and rising per capita income, while agriculture's natural resource base will become increasingly challenged. A fundamental issue for the future is to produce more with less while sustaining and increasing the livelihood of family and small-scale farmers. To fulfill expanding and changing food demand while also halting and reversing environmental deterioration, significant advancements in resource-use efficiency and increases in resource conservation will be required globally. Despite certain technical advancements, crop growth has slowed dramatically compared to earlier decades. The detrimental side effects of excessive usage of chemical inputs in agricultural production are becoming increasingly obvious, raising severe questions about long-term sustainability.

Figure 6.10: By 2050, food consumption is anticipated to rise anywhere from 59 percent to 98 percent. This will have a profound impact on agricultural markets that we have never seen before. Farmers throughout the world will need to improve crop output, either by expanding the quantity of agricultural area available for crop production or by improving productivity on existing agricultural lands with fertilizer and irrigation, as well as embracing innovative technologies like precision farming.

Source: https://hbr.org/2016/04/global-demand-for-food-is-rising-can-we-meet-it

Agriculture, fisheries, and forestry investments, as well as research and development funding, must be increased, particularly in and for low-income nations. This is necessary to encourage the adoption of sustainable production systems and practices, such as incorporated crop-livestock and aquaculture-crop systems, agroforestry, conservation agriculture, sustainable forest management, nutrition-sensitive agriculture, and sustainable fisheries management, among others. These and other types of climate-smart agriculture will aid farms, ecosystems, and communities in adapting to, mitigating, and building resilience to climate change, as well as meet country-specific demands and gender issues (Blissett & Fogel, 2013). Furthermore, because current pricing incentives and supports frequently work against sustainable agriculture, agricultural subsidies, both implicit and explicit, must be realigned.

6.3.2 Availability of Sustainable Natural Resources

According to projections, agricultural land, forests, water, biodiversity, and capture fisheries will face increasing pressures by 2050. The additional land required for agricultural production between now and 2050 is anticipated to be little about 100 million hectares. Demand for agricultural land is predicted to fall in high-income nations while rising in low-income ones. The minor rise that resulted might indicate that land availability is not a restriction. In actuality, expansions in agricultural land are hampered by the fact that arable land is inaccessible owing to a lack of infrastructure, geographic isolation from markets, or disease outbreaks, or is prone to disease outbreaks. Furthermore, the land that is accessible is concentrated in a few countries (Capitanio et al., 2010).

Water scarcity will also become a rising issue, particularly in locations where a large amount of their water resources is used and when production systems are subjected to high levels of environmental and social stress. Water shortage has ramifications for women's capacity to access productive resources, in addition to restricting the possibilities for developing irrigated regions. The rate of growth of irrigated land is already decreasing significantly. Future water stress will be caused not just by increases in demand, but also by fluctuations in the availability of water resources as a result of climate change-induced changes in precipitation and temperature.

6.3.3 Climate Change

Damage and losses in production; degradation of water, forests, land, fish stocks, and other natural resources; decreasing rates of productivity growth; and added increased pressure on already fragile agricultural ecosystems and livelihoods are all concerns raised by global warming and natural and human-caused disasters. Maintaining the planet's natural resource base's capacity to feed the world's rising population while decreasing agriculture's environmental and climate imprint is critical for current and future generations' well-being. Climate change consequences will put food security and human livelihoods at jeopardy far beyond 2030. Climate change has an influence on food availability, agricultural yields, seafood populations, and animal health. It restricts food access by having a detrimental impact on rural livelihoods and incomes. Global warming is also seen as a key 'risk multiplier' for hunger. According to some estimates, climate change will put an extra 120 million people at risk of malnutrition by 2050, with 24 million of them being children; over half of the rise would be focused in Sub-Saharan Africa (Chaudhry et al., 2008).

Until 2030, the negative effects of climate change will only marginally outnumber the good effects. Increased plant growth in warmer temperatures will mostly benefit temperate zones at higher latitudes, whereas negative consequences will primarily affect tropical zones at lower latitudes. After 2030, negative effects will become more severe, with large yield losses in most regions of the world no longer being offset by positive yield improvements in other places. Droughts and floods, for example, will worsen and become more common as a result of climate change. Climate change may have an influence on nutritional outcomes by altering the micronutrient content of specific foods and compromising food safety. Furthermore, extreme temperatures and weather extremes promote the growth of food-borne infections, which impair the body's ability to utilize nutrients.

Figure 6.11: Climate change is accelerating at an extraordinary speed. A mountain of data implies that global temperatures will remain to rise, and that human activities like fossil fuel burning, deforestation, and agriculture are the primary contributors. These activities emit greenhouse gases (GHGs) such carbon dioxide, methane, and nitrous oxide, which trap the sun's heat and warm the atmosphere, hence the term "greenhouse." Natural processes also emit GHGs; however, the ability of trees, soil, seas, and other sinks (storehouses) to sequester (catch and store) emissions has typically counterbalanced these emissions.

Source: https://www.foodsystemprimer.org/food-production/food-and-climate-change/

6.3.4 Poverty and Inequality

Despite worldwide economic progress and poverty reduction during the previous 30 years, around 2.1 billion people remain poor, with 700 million living in severe poverty. High and growing inequality is obstructing efforts toward poverty elimination. Even in nations where poverty has decreased, there are still significant disparities between urban and rural areas, regions, ethnic groups, and men and women. The majority of the world's impoverished and hungry live in rural areas, where agriculture, fishing, and forestry provide a meager living. Agriculture is the key to poverty and hunger reduction since poor people rely on agriculture for their livelihoods and spend a large portion of their family budgets on food. Agriculture's structural change has stopped in areas where income growth has been reduced, leaving many people in poverty (Jankovic et al., 2010).

Agriculture is critical to pro-poor development. Increasing production and profitability, connecting farmers to markets, and providing effective extension and agricultural advising services are all necessary steps in reducing rural poverty. However, factors other than agriculture have a role in pro-poor growth. It necessitates high-quality education, economic diversification into non-farm income-generating industries in rural areas, job creation assistance, and suitable social safety systems.

Women have more hurdles to productive resources, economic opportunity, and decision-making than males everywhere. Agricultural productivity development and agricultural production are hampered for farming women due to a lack of access to agricultural supplies, services, loans, and markets, making the challenging route out of poverty even more difficult. Because of the gender difference in access to resources, female agricultural laborers in Sub-Saharan Africa have productivity levels that are 20 to 30 percent less than male employees.

To end severe poverty and hunger, significant extra expenditures are required. However, due to low present levels of capital creation and restricted 'fiscal space' in low-income countries, there would be a need for international financial cooperation to finance investment programs. Future population growth will be disproportionately concentrated in nations with significant levels of food insecurity. While rising average per capita income is predicted to improve nutritional outcomes, tackling the triple burden of malnutrition — undernutrition, micronutrient deficiencies, and obesity – will continue to be a problem for those nations in the coming decades (Caporgno & Mathys, 2018).

In low-income nations, population growth is predicted to drive up demand for basic crops such roots, tubers, and plantains. At the same time, changes in dietary patterns will be driven by income development and urbanization, with significant rises in need for cereals, milk, and meat products. Increased intake of animal products and high-fat, high-sugar diets, along with urban inactive lifestyles, will raise the risk of obesity and overweight.

Improving vulnerable people' access to food and guaranteeing urban food security, particularly in low- and middle-income nations in Asia, Africa, Latin America, and the Caribbean, will be critical to ending hunger in the next 15 years. Agriculture and food systems will have to accommodate the food and nutritional needs of individuals with income growth and changing diets, as well as a growing number of impoverished and hungry people (Chen et al., 2013). While increasing farm production to meet this demand has received a lot of attention, supply chains that connect farmers to urban markets, as well as policies like pricing and social protection, are equally important in ensuring that consumers have access to nutritious and safe food at affordable prices.

In regards of greenhouse gas emissions and natural resource usage, the shift in food patterns will have a higher environmental impact. Increased emissions of methane from anaerobic fermentation, carbon dioxide from destruction of trees for pasture, and nitrogen oxides from feed production are all linked to the shift to diets heavy in milk and meat, notably from ruminants. Increased use of processed foods necessitates increased usage of water and energy, both of which have significant environmental consequences if these resources are not handled responsibly.

Although additional study is needed, data shows that low-environmental-impact food habits can also be associated with excellent health. National dietary guidelines, for example, that encourage reducing red meat intake, especially among high-consumption groups, might help reduce GHG emissions.

6.3.5 Inclusivity, Efficiency, and Resilience

The coexistence of contemporary and traditional supply routes is a feature of food systems. However, as many areas become more reliant on global supply chains and massive distribution networks, such as supermarkets, the systems are evolving. Vertically integrated, capital-intensive supply chains adapt to changing needs for food and food habits while also shaping their evolution. The high-calorie, low-nutrient content of many foods; the reduced access of local suppliers and family farmers to viable markets; high levels of food wastage; food safety issues; plant and animal medical concerns; and the higher energy severity and heavier carbon impact associated with the prolongation of food chains are all new challenges and concerns (Hirschi, 2012). The consequences of these problems for future food security and nutrition must be considered in the context of entire food systems, including the effects on conventional food chains and the farmers and consumers that rely on them.

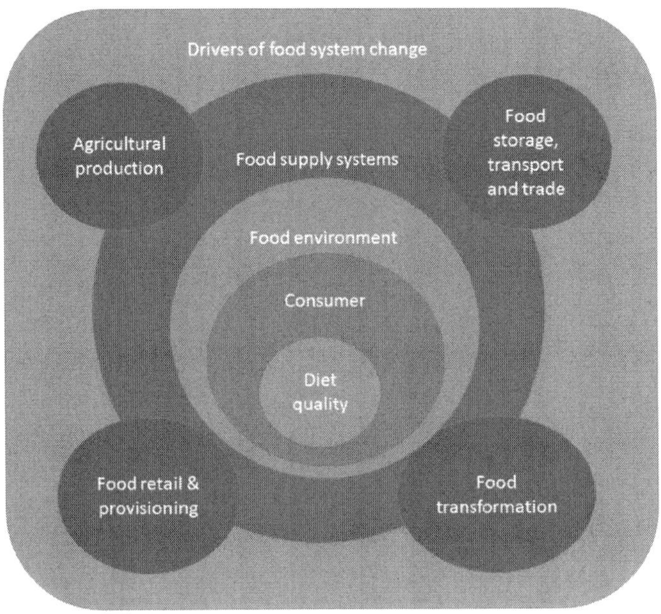

Figure 6.12: In reaction to socioeconomic and demographic shifts, food systems in low- and medium-income nations are undergoing a fundamental transformation. Rising wages, rising urbanization, and the rise of the middle class all result in significant changes in dietary habits and consumer behavior, necessitating governmental and private efforts to promote food market integration. Increased agricultural production has only partially translated into reduced hunger, and it has not been matched by a drop in malnutrition.

Source: https://ecdpm.org/great-insights/sustainable-food-systems/challenges-perspectives-sustainable-inclusive-food-systems/

In both rural and urban locations, strengthened links between farmers, markets, and consumers may be a key source of income development and employment creation. Formal, organized supply chains improve product flow efficiency – from inputs to farmers and food commodities to consumers – but they have also been discovered to constitute a threat to food security. Distribution networks, for example, may be focused in more affluent urban regions. Furthermore, small manufacturers may find it difficult to match the expectations of huge supermarkets for uniformity, regularity, regular supply, and high volume. Concerns concerning efficiency and equality are being raised as a result of the effects of structured supply chains. Despite the 'supermarket revolution' and the emergence of contemporary global food supply networks, local food systems remain essential.

Food losses occur across the food value chain in low-income nations due to management and technological constraints in harvesting, storage, transportation, processing, packaging, and marketing. Food waste is mostly driven by consumer action and policies and laws that target other sectoral objectives in medium and high-income nations. Subsidies, for example, may encourage surplus food agricultural production, lowering both prices and the amount of attention paid to food wastage. Some food quality and safety regulations may cause food that is still acceptable for human consumption to be removed from the supply chain. Inadequate purchase planning and failure to eat food before it expires can contribute to waste at the consumer level (Han et al., 2018).

Finding dynamic paths that connect sustainable agricultural practices to rising urban markets and seizing market possibilities will be a problem for many low- and middle-income nations. The majority of demand for high-value foods, like fruits, vegetables, and dairy products, comes from cities. Because such items are labor-intensive, small-scale and family farmers may have an edge in these marketplaces. Rural poverty reduction and agricultural growth may be greatly aided by food networks that connect farmers to cities. Connecting small-scale farmers with supermarket distribution networks through mutually advantageous contractual arrangements and giving new priority to the growth of local food systems are two options.

6.3.6 Earning Opportunities

Inequalities are pervasive and persistent, trapping far too many rural residents in hunger and poverty. Young people in low-income nations' rural areas are generally hesitant to engage in low-productivity agriculture. They join the flow of international and internal migrants in the absence of adequate economic prospects and access to social services and protection. Women and elderly people are often entrusted to care for the farm in many places, yet they confront significant challenges in gaining access to resources to boost their production. Addressing those imbalances through more inclusive rural reforms and reconfigured rural-urban links will be a key issue in the next decades.

Thousands of millions of young people will need to be integrated into the labor market in the next decades, which is arguably the single largest global development problem. By 2050, the number of persons aged 15 to 24 will have increased from 1 billion in 2015 to 1.2 billion. The majority of these youth will reside in South Asia and Sub-Saharan Africa. In rural regions, high levels of young unemployment and underemployment prohibit households from diversifying their livelihoods and permanently escaping poverty (Granato et al., 2010). Poverty reduction requires the development of human capital via the provision of high-quality basic social services, notably in the areas of education and health. Nevertheless, in many low- and middle-income nations, population expansion is exceeding the creation of new jobs, and rising urbanization has not been matched by an increase in non-agricultural employment. As a result, agricultural and agriculturally based services will continue to process a significant number of new workers.

Migration is an element of economic growth and agricultural structural transformation. Population growth, industrialization, climate change, and political strife will all drive distress migration in the next decades, both within and beyond nations. Another important new problem is managing migratory flows, which must be addressed by addressing the core reasons and enhancing access to social security and economic possibilities in both destination and origin nations.

6.3.7 Disasters, Conflicts, and Crisis

Hunger, malnutrition, and poverty are most difficult to combat in nations that have been through long-term crises. The causes of these crises are a mix of recurring factors, including human-made and natural hazards that often occur at the same time, violent conflict, long periods of food scarcity, the rundown of livelihoods and food systems, and insufficient governance and institutional capacity to cope with the consequences. Prolonged crisis conditions affect about half a billion individuals in more than 20 nations and territories, predominantly in Africa. Agriculture and allied sectors provide the majority of these people with food, income, and well-being. Two-thirds of global humanitarian aid, or 80% of OECD member nations' emergency budgets, has gone to alleviate chronic or repeated crisis situations, which persist on average eight years or more (Granato et al., 2010).

Conflicts, as well as long-running crises and natural catastrophes, are severe hindrances to farm livelihoods, food security, and nutrition. Displacement and migration fluxes are also fueled by them. Conflicts and disasters have been more intense and frequent in recent decades across the world. To avoid and resolve growing conflicts throughout the world, more risk-informed, inclusive, and equitable resiliency and development approaches will be required.

6.3.8 Food Systems Threats

The quantity and severity of outbreaks of intergovernmental plants and animal's pests and illnesses is increasing at an alarming rate in agriculture. Food safety is a concern for food systems in general, as is the possibility of radioactive incidents. The growth of food system emergencies is partly due to climate change.

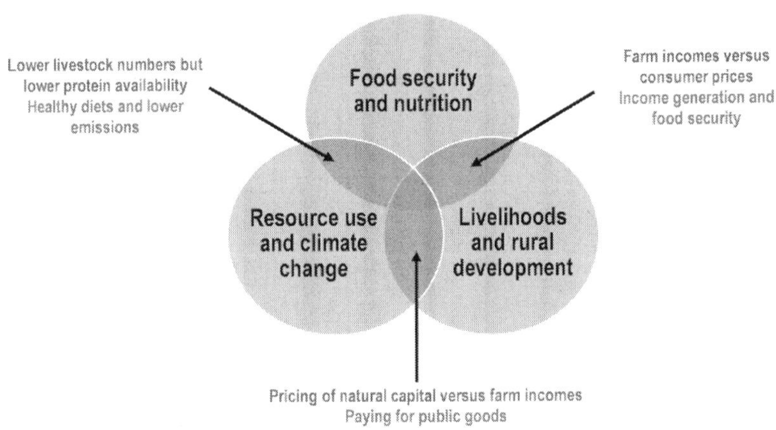

Figure 6.13: Food production also has a significant environmental impact. Agricultural production (encompassing pre-production and post-production operations) accounts for 21-37 percent of anthropogenic greenhouse gas emissions; almost 80% of all vulnerable terrestrial bird and animal species are in risk owing to habitat loss due to agricultural growth.

Source: https://www.oecd-ilibrary.org/sites/edf73cce-en/index.html?itemId=/content/component/edf73cce-en

Controlling transboundary plant diseases and pests enhances production and decreases yield losses in crops and pastures. Integrated pest management, which favors biocontrol and biopesticides that safely reduce pest and disease threats, can help achieve this. Animals die and become ill at a high rate as a result of transboundary animal illnesses. They continue to disrupt regional and international livestock markets and commerce, putting livestock producers' livelihoods at risk all around the world. The international community now lacks the ability and cooperation needed to prevent, manage, and eliminate new transboundary animal illnesses. Intensive animal production to meet shifting food demand increases the danger of point-source contamination, increased antibiotic usage, and possibly more catastrophic zoonotic disease outbreaks.

Food-borne illnesses are a leading source of morbidity and death across the world. The scope and expense of hazardous food, as well as the burden posed by parasite and chemical contaminants in food, are, nevertheless, largely understood. Unsafe water used in food production, filthy food handling, insufficient storage facilities, and poorly implemented rules may threaten food safety even further. Antimicrobial resistance is on the rise, posing a danger to the treatment and management of a wide spectrum of illnesses. Antimicrobials are still widely utilized to safeguard human and animal health, as well as livestock and agricultural productivity in general (Gaudin, 2017).

6.3.9 International and National Governance

Given the interwoven nature of the issues confronting food and agriculture, tackling them will need coordinated policy responses at both the national and international levels. Given the poor track record of sector-specific policymaking and the gaps in global and national governance processes, regulatory systems, and surveillance and accountability, developing such methods will be difficult.

The 2030 Agenda for Sustainable Development, as well as associated global accords, emphasize the interconnection of the issues that the world faces on the road to sustainable development. They understand that multiple activities must be combined to accomplish connected goals, and that this integration will impose new technical requirements on policymakers at all levels, as well as new demands on institutional structures and coordination at all levels of governance. Combining instruments implemented at different levels of governance in ways that are mutually reinforcing while comprising unavoidable trade-offs; and commercializing on synergies among the Sustainable Development Goals (SDGs) and key initiatives, among different sectoral policies, and among diverse stakeholders at international, regional, national, provincial, municipal, and local, levels are among the related challenges (Dobrucka & Cierpiszewski, 2014).

To strengthen communication about the difficult policy decisions that must be taken, more equitable governance will be required. Avoiding the exclusion of the poor, who lack the legislative power to influence choices, and gradually including them in the developmental process is critical. Growing competition for natural resources has the potential to deprive the rural poor of the basic foundations of their livelihoods, particularly in long-term crisis circumstances, conflict-affected regions, and disaster-affected areas. Ensure that the poor's official and informal rights to access and exploit natural resources are recognized through the application of cooperative guidelines on the sustainable governance of land, fisheries, and forests, as well as assistance for the fulfilment of the right to adequate nourishment.

The importance of international collaboration cannot be overstated. Many of the resources that the agriculture industries rely on are transboundary in nature. Modifications in the environment will affect resource availability, human migration, and animals and plants species movement, as well as human activity. Extreme occurrences such as wildfires, species invasions, pests and illnesses will transcend national borders. Institutions and policies dedicated to the prevention and administration of specific climate-related hazards and vulnerabilities are mostly local and national, but international collaboration and procedures are needed to support them more effectively (Costa & Jongen, 2006).

Financing for inclusive agriculture and food development; addressing employment and migration challenges; attempting to address shortfalls in the international monetary regime in relation to agriculture and food systems; and supplying open access to information and statistical data to strengthen the role of all interested parties in governance are some of the other areas that require improved governance.

6.4 CONCLUSION

In product development, best practice is a moving objective. Not only are new practices being established and perfected, but the diversity of organizations necessitates the use of these practices in a customized manner. In product development, there are eight basic concepts and four basic stages that apply to all firms, projects, and periods. However, as the company's ideology, knowledge, abilities, and assets evolve, so do the sorts of product innovations and product development operations. Product creation is a critical strategic problem that requires continual attention, according to successful firms.

Product development performance and success rate (product development effectiveness and efficiency) must be evaluated, and this assessment must be combined with the company's strategic goal in order to define and organize future improvements in both effectiveness and efficiency. Because creativity and criticism are two competing thinking processes, this is not an easy task (Dimkpa & Bindraban, 2017). Creativity, which is critical to product creation, ventures into the unknown and makes mistakes; product development assessment seeks out and criticizes these errors. Focusing on faults leads to cautious product development; focusing on creativity leads to wild product development; effective businesses combine creativity and assessment in their projects.

Product development is unique to the firm and is linked to its history, philosophy, and knowledge, but the organization's position in relation to industry and market best practices is an indicator of its ancient and modern product development efficacy and efficiency. Strategic plans for enhancing product development can be developed based on the results of this review. The efficacy and efficiency of product development are combined to provide the strategic product success that the company's senior management has predicted. The efficacy and efficiency of product development are tightly related.

CHAPTER 7

DRIVERS OF NEW FOOD PRODUCTS DEVELOPMENT

CONTENTS

7.1 OVERWEIGHT/OBESITY E.G. LOW-FAT YOGURT

Current production and consumption patterns, as well as newly emerging eating habits, have an impact on health, the environment, and society. Obesity, osteoporosis, cancer, diabetes, allergies, and dental problems are just a few of today's modern ailments.

Obesity is on the rise around the world, and nutrition and exercise are critical for preventing and treating it. Obesity and weight loss may benefit from nutraceutical (natural product) therapy.

Nutritional research has discovered several substances that can help with weight loss, including carbs with a low glycemic index and 5-hydroxytryptophan (da Silva Vaz et al., 2016). Throughout 24 hours, chromium, green tea, and the first two nutrients all work to suppress hunger while increasing energy expenditure.

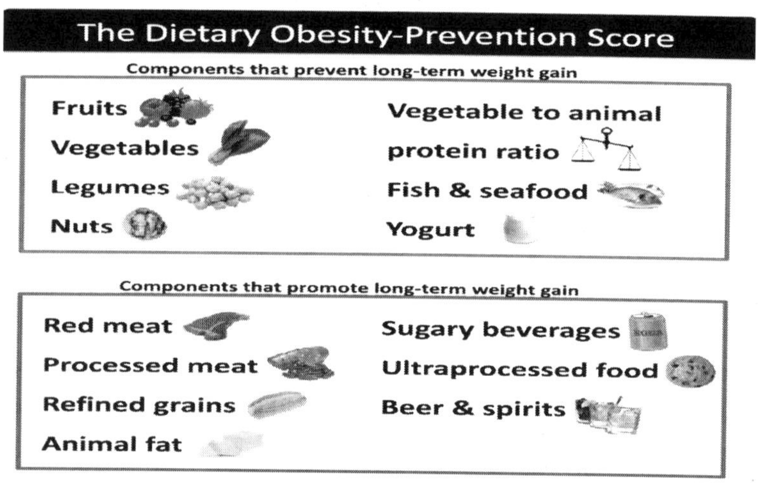

Figure 7.1: A food-based score and incidence of overweight/obesity.

Source: *https://els-jbs-prod-cdn.jbs.elsevierhealth.com/cms/attachment/759d853a-f1c4-40dc-93da-2c8aefd3abe5/fx1_lrg.jpg*

Obesity and weight gain are the world's sixth-leading causes of death. Every year, at least 2.8 million people die as a result of obesity-related causes. Obesity and excessive weight gain are responsible for 44 per cent of diabetes cases, 23 per cent of ischemic heart disease cases, and 7 per cent to 41 per cent of certain cancer cases. As a result of their ageing populations, developed countries are also confronted with issues such as high-energy meals and unbalanced diets.

Obesity is on the rise around the world, and nutrition and exercise are critical for preventing and controlling it. Obesity and weight loss may benefit from nutraceutical (natural product) therapy. Nutraceuticals, or health-enhancing foods, are food-like substances that alter the physiological activities of the body (da Silva Vaz et al., 2016). Furthermore, items with this function are sold and consumed for their added value, which is why this function is so important. Obesity has been elevated to prominence in food policy and commercial strategy as a result of a worldwide convergence of public events. The result has been a trend toward low-carbohydrate eating, but the real potential lies in the long-term development of foods with functional effects on obesity — or functional foods for obesity therapy.

Nutraceuticals, also known as "functional foods," have sparked heated debate because they blur the line between food and medicine. "Functional food" is food that is cooked or prepared with or without a knowledge of how or why it is used. As a result, a functional meal contains all of the nutrients you require to be healthy and strong. Nutraceuticals are dietary supplements that aid in the prevention and/or treatment of diseases and disorders other than nutritional deficiencies like anemia. This means that what one person considers to be a functional meal may be a nutraceutical to another. Fortified dairy products (milk is a nutrient because one of its natural elements, casein, is a pharmaceutical) and citrus fruits (orange juice is also a nutrient because its constituent ascorbic acid is a pharmaceutical) are examples of nutraceuticals.

7.2 WHERE ARE WE NOW WITH OBESITY AND NUTRACEUTICALS?

Diet and exercise are critical in the prevention and treatment of obesity, a worldwide problem. Conjugated linoleic acid (CLA), capsaicin, Momordica charantia (MC), and psyllium fibre are examples of antiobesity nutraceuticals. In obese people, glucomannan, chitosan, fenugreek, G. Sylvestre, and vitamin C were found to reduce body weight and increase fat loss. This treatment's long-term effectiveness and potential side effects must be investigated further.

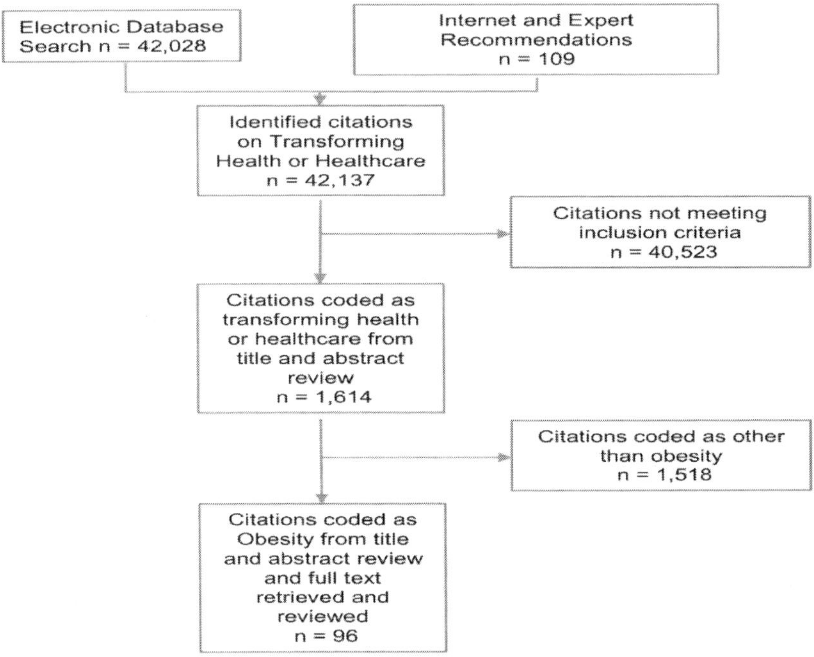

Figure 7.2: Systematic review into city interventions to address obesity.

Source:________________https://els-jbs-prod-cdn.jbs.elsevierhealth.com/cms/attachment/841deb85-edbc-4cbf-a6f6-cd7726b7000c/gr1_lrg.jpg

7.3 OBESITY

Obesity is the leading cause of long-term health problems and disability worldwide. On a worldwide scale, more than a billion people are overweight, with at least 300 million clinically obese. Obesity affects people of all ages and income levels, and it is a complex disease with significant social and psychological consequences. It is frequently found alongside malnutrition in developing countries. Between 1980 and 2008, the global obesity rate nearly doubled. In 2008, national health authorities in the WHO European Region classified roughly half of all men and half of all women as overweight or obese. Given the rising prevalence of obesity around the world, it is critical to developing effective techniques for preventing and treating obesity. Any weight-reduction regimen should result in a 7–10% weight loss from the start.

Obesity occurs when a person's daily caloric intake exceeds their daily caloric expenditure, resulting from a metabolic imbalance. One or both of the two components of the body's energy balance must be adjusted to successfully treat or prevent obesity. As a result, weight reduction treatments (including functional eating strategies) can target various parts of the energy balance system, including food intake, energy expenditure, and energy. Because 61 per cent of the population is overweight or obese, pharmaceutical companies are looking into all of these solutions (Yang et al., 2017). Producing weight reduction functional meals, on the other hand, maybe a more enticing strategy.

The goal of this essay is to provide light on the reasoning behind the development of innovative food products to supplement current medications. Each component of the product is listed in order of relevance based on clinical effectiveness.

7.3.1 Eating Functional Meals May Allow You to Conserve Energy

Functional meals can assist you in losing weight by boosting your sensation of fullness for a longer period. Providing meals that increase a person's perception of fullness and encourage them to stop eating early can assist achieve the goal of lowering overall energy intake. Three of the most promising areas for improving satiety include adjusting the meal's caloric density, macronutrient composition, and glycemic index.

The amount of energy contained in a unit of weight or volume is referred to as dietary energy density, and it appears to be related to overall energy consumption. Calculating the energy density of most meals is as simple as multiplying the kilojoule (kJ) value by the weight or volume (g) of the food. Dietary energy density is substantially related to energy intake in the short term.

Depending on the total weight or volume consumed, food consumption may serve as a physiological signal of fullness. Thus, changing one's diet to lower total calorie consumption and, as a result, obesity may be advantageous. The non-caloric components of the meal, particularly the water content, have the largest influence on energy density. Those with a high water content have a lower energy density than foods with low water content.

As a result, the energy density of a meal decreases since fibre adds more to the food's weight than calories. Several studies have shown that high-fibre diets had lower total caloric consumption than low-fibre diets, indicating that dietary fibre intake has the greatest influence on overall energy consumption. For several reasons, high-fibre diets are associated with lower food consumption (Wyrwa & Barska, 2017). To begin, a high-fibre diet may increase the desire to chew. Fibre-rich diets may cause a delay in stomach emptying and nutrition absorption. Diets high in fibre lower the total calorie density of meals, making them less calorific overall. Several research shows that increasing dietary fibre may aid in weight loss.

The amount of energy in a meal is determined by its macronutrient composition. Reduced-fat consumption in the diet can lower total energy density since fat has a higher energy density (38 kJ/g) than either protein or carbs (17 kJ/g). Functional meals with reduced calorie density may improve obesity treatment since there is substantial evidence that doing so decreases energy intake. As a result, further long-term research on the impacts of energy density and how it affects body weight is required. Dietary fat type alteration as a weight reduction method is infrequent (Wenzl et al., 2006). According to several research, the concept that polyunsaturated fatty acids increase total fat burning more than saturated fatty acids is debatable. While short- and medium-chain triacylglycerols (SCTG) have been demonstrated to boost energy metabolism, it is unknown if they will help with weight loss.

The rise in blood glucose following ingestion is compared to the rise in blood glucose following a regular meal, such as white bread, to calculate a food's glycemic index. Increased insulin production after eating can result in increased hunger and a variety of insulin-related illnesses. The effect of a meal's glycemic index on calorie consumption and obesity is being debated at the moment. Although there is no solid evidence that food consumption directly impacts the glycemic index, it has been connected to weight gain. Consuming items with a low glycemic index can help to lower a diet's total glycemic index. If the glycemic index affects food consumption, creating a more pleasant, low-glycemic meal is a positive goal.

7.3.2 Functional Meals Help to Meet Your Body's Energy Requirements

Another method for preventing or treating obesity is to improve your overall energy expenditure while decreasing your calorie intake.

While some dietary supplements claim to increase energy expenditure, the evidence to support these claims is lacking.

7.3.3 Caffeine and Ephedrine are not Stimulants

Caffeine and ephedrine have been demonstrated to be helpful for long-term weight control, most likely owing to the combination of various mechanisms, such as lowering phosphodiesterase-induced cAMP breakdown and increasing sympathetic catecholamine release. However, fresh research has raised concerns regarding ephedrine's long-term consequences.

Green tea contains catechins and caffeine, both of which have been shown to inhibit catechol O-methyltransferase and phosphodiesterase, respectively. In this instance, the mechanisms may cooperate. Tea catechins also have antiangiogenic characteristics, which may help prevent obesity. It is also plausible that sympathetic nervous system-stimulated white adipose tissue plays a significant role in total body fat management.

Because of its high catechin concentration, oolong tea may also help you burn more calories. After three days of daily administration of five cups of oolong tea, the resting metabolic rate rose by 3–4%. The majority of the increase in metabolic rate, which has the biggest impact on fat loss, was unexpectedly accounted for by cells' increased capacity to oxidize fat. It has been proposed that the flavonoid content of green tea, particularly its polyphenolic level, might be employed to increase energy expenditure. Catechin O-methyltransferase (COMT), which is inhibited by catechins, prevents norepinephrine breakdown. Epigallocatechin gallate (EGCG) is the most prevalent catechin found in green tea. This is not found in significant quantities in any other dietary source. Catechins block COMT, extending norepinephrine's thermogenic and fat-burning actions (Vidović et al., 2022). Through the sympathetic nervous system, norepinephrine regulates both of these metabolic processes. Because norepinephrine has a longer half-life, it may linger in the sympathetic synaptic cleft and exert its action for a longer period. Caffeine elevates norepinephrine levels by inhibiting phosphodiesterases and extending the half-life of cAMP in the cell. Thermogenesis is greatly impacted by these activities and EGCG's long-lasting norepinephrine effect.

Even though capsaicin has been demonstrated to be beneficial, clinical usage necessitates strict adherence to the recommended dose. There is currently no proof that this dose is safe. A recent study suggests that capsaicin may help with weight reduction and fat growth by affecting protein changes in the body.

According to a new study, calcium-rich meals have been linked to an increase in energy expenditure, which may aid in weight loss. Increasing calcium consumption has been related to a reduced BMI in various studies, although there is no evidence that this is a direct relationship.

7.4 MORE MEDICINES CAN AID WITH WEIGHT LOSS

Brain serotonin concentrations

Serotonin reduces the brain's tendency to overeat. According to this study, amino acids may help in weight loss. When tryptophan is converted to 5-HTP, serotonin is produced. Serotonin appears to control both energy balance and eating habits daily (i.e. three times a day) through activating satiety neurons in the medial hypothalamus (Vigani et al., 2015). To lower appetite and carbohydrate consumption, serotonin and norepinephrine appear to interact antagonistically. This dose had no negative side effects, and no one withdrew due to any documented symptoms.

Chrome consumption

Dietary chromium is a trace element that helps to increase insulin action.

Glucose intolerance, insulin resistance, and increased blood glucose levels are all symptoms of chromium insufficiency. The use of supplements may assist to reduce the intensity of these symptoms.Weight reduction may occur from improved insulin usage. Insulin resistance treatment can increase muscle mass by decreasing catabolism.

Chromium picolinate has no impact on weight reduction. Rather, it appears to alter the fat-free mass-to-fat-free mass weight loss trajectory. Doses of 200 or 400 g appear to be effective, with chromium picolinate appearing to be superior to yeast bound chromium.

Developing a useful food product

By integrating scientifically proven bioactive components in a dietary supplement, scientists pioneered an entirely new approach to weight control with this product.

Carbohydrate/Fiber Sources

This snack contains fructose, barley, and psyllium fibre, all of which have a low glycemic index. These meals increase satiety and have been linked to weight loss and a decrease in co-morbid disorders.

Fructose is the primary sweetening component of this product. This is the kind of sugar found in the majority of fruits. Fructose is a sweeter sugar than sucrose, which is usually made from beet or cane sugars. It has a low glycemic index ([GI] = 32). Fructose has a lower incremental influence on plasma glucose levels when consumed as part of a meal than sucrose, glucose, potato starch, or wheat starch (Abbas et al., 2010). The functional food includes three grams of fructose per serving.

Psyllium husk fibre is a viscous, water-soluble fibre derived from the seeds of the psyllium plant. As a dietary fibre, psyllium promotes satiety and assists in weight reduction. It has also been demonstrated to lower blood cholesterol and glucose levels. Each serving of this product contains 1.8 g of sugar.

Ceryain barley has a low glycemic index, making it one of the few grains. Sweet maize (GI = 78), rolled barley (GI = 94), and instant white rice (GI = 128) all had lower glycemic indices than pearled barley (GI = 36) and cracked barley (GI = 72). Barley is a low-glycemic carbohydrate with a high fibre content (1.5%), which aids with blood sugar regulation and weight management. Each serving contains one gram of sugar.

Additional Ingredients Sources

Two servings of this functional food product contain the recommended daily dose of green tea extract, which has been shown to help with weight loss. Each serving includes 135 mg of EGCG and 75 mg of caffeine. While some patients may have reservations about drinking coffee to help them lose weight, these reservations are unfounded. To begin, two servings of the product have 150 mg of caffeine in them. In contrast to other foods, this is a decent amount. A cup of coffee has around 180 mg of caffeine, whereas a 12-ounce can of cola contains 64 mg. An OTC pain reliever (Excedrin) normally includes 130 mg, but a weight-loss supplement (Dexatrim) contains 200 mg. Second, due to caffeine's short half-life of 3 to 3.5 hours, it has not been proven that residual effects from 100 mg dosages taken during the day have an impact at night (Veeck, 2010).

The product includes just ten milligrams of 5-hydroxytryptophan per serving, which is far less than the clinically effective amount of 450 mg per serving. Each meal contains 100 micrograms of chromium picolinate. This is the smallest effective dose range. Individuals who used the product were expected to acquire an additional 200 mcg of vitamin D through their regular diet, putting them at the upper end of the recommended dose range.

Obesity has reached epidemic proportions around the world, impacting nearly the same number of adults and children. Despite millions of dollars spent on weight control, only a few therapies have been proven to be successful. Patients and practitioners alike value breakthrough products that can help curb the outbreak. Because of the present level of knowledge on functional foods, a dietary supplement comprising active molecules that enhance satiety, boost calorie and fat metabolism, and promote fat tissue loss rather than lean tissue loss has been developed (Van Huis, 2020). Each active component is sourced from a food that has been clinically proven to work. Active chemicals were combined in the right amounts and sourced from the same sources as those evaluated in clinical studies to assure effectiveness.

7.5 CANCERS

Cancer is the world's second greatest cause of death, after only cardiovascular disease. The body's cells can grow uncontrolled and hierarchically, killing and replicating normal tissue. This is an inherited disorder. Carcinogens and random mutations are the two most prevalent causes of these changes. Carcinogens are substances or chemicals that can alter the DNA and cause cancer. Some interact directly with DNA, whereas others interact with chemicals capable of forming covalent adducts with DNA in the presence of stimuli. Mutations in crucial genes wreak havoc across the body. Carcinogens can be produced in the laboratory or the environment. This technique can recognize cancer-causing substances from both external and internal sources. Exogenous carcinogens that are biological, physical, or chemical all-cause DNA changes when they enter the cell via membrane receptors or passive diffusion. A range of carcinogens, such as ultraviolet (UV), ionizing, X-ray, and radioactive radiation, as well as oestrogens and prolactin in smoke, and biological carcinogens such as DNA and RNA, as well as several viruses, damage DNA while simultaneously stimulating cell growth and differentiation.

Endogenous substances impact humans in ways that metabolic byproducts such as reactive oxygen species (ROS) do not (ROS). ROS-induced DNA replication mistakes, according to Badger, might be a sign that a cell's genetic code has been seriously damaged. ROS-induced oxidative stress has been found to alter tumour development and carcinogenesis pathways.

Food, an external carcinogen, is one of the world's top causes of cancer today. Heat treatment, cooking processes, and the eating of genetically engineered foods can all produce a variety of harmful chemicals. These elements interfere with DNA replication, which contributes to cancer. The International Agency for Research on Cancer (IARC) assessed and quantified carcinogenic risk in over 1,000 chemical agents and combinations linked to an increased risk of cancer in humans. It is crucial to note, however, that your dietary choices have a substantial impact on your risk of cancer (Tarasuk, 2008). According to studies, many individuals consume hormone foods (27%), burnt meals (14%), alcohol (16%), and cola beverages (7%), all of which are known to cause cancer. The discovery of possibly carcinogenic chemical compounds in food raised concerns about the safety of heated meals and beverages.

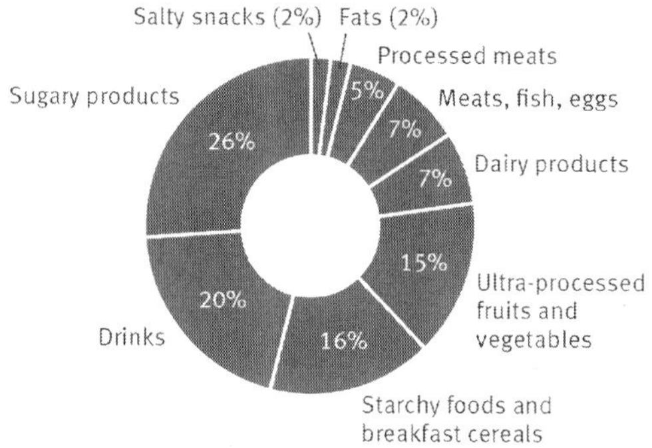

Figure 7.3: Consumption of ultra-processed foods and cancer risk.

Source: https://www.bmj.com/content/bmj/360/bmj.k322/F1.large.jpg

Food carcinogens and dietary habits have been connected to cancer development.

Nutrition is connected to 70% of all malignancies and 40% of cancer-related deaths. Several studies have been conducted to investigate the potentially hazardous consequences of certain foods, chemicals, and contaminants. According to cancer risk research, a poor diet and sedentary lifestyle are connected to breast cancer recurrence and mortality in Mediterranean nations. This trend is caused by increased life expectancy and Western-style lifestyles, which are characterized by various risk factors, both non-modifiable (age and genetic disposition) and modifiable. Adipose tissue inflammation promotes the formation and spread of breast cancer and is associated with obesity, which is caused by a high caloric intake of sugar and saturated fat. Red meat consumption has long been suspected of being a cause of cancer. Cooking processes and high temperatures are two elements that lead to the formation of pro-carcinogenic chemicals in meat.

Protein denaturation and solubility changes, as well as vitamin loss and fatty acid oxidation, can all lead to decreased product quality. Saturated fatty acids, particularly those found in dairy products, have been related to an increase in breast cancer incidence as well as an increase in the production of endogenous insulin-like growth factor (IGF-1) in breast cancer growth (IGF-1). As a result of the hazardous manufacturing process, salted fish contains a variety of possible carcinogens, including N-nitrosodimethylamine. Fermented or distilled alcoholic drinks can cause cancer in humans and accelerate tumor formation.

To put it another way, food is a major cause of cancer in these situations due to the death of taste receptor cells (TRCs), which causes a loss of satiety, nausea, and vomiting. As a result of increased protein catabolism and energy expenditure, cachexia and weight loss have been described. Smoking, drinking, and food consumption all have a direct detrimental influence on the mouth cavity, which acts as an access point to the digestive system, and as a result, several disorders and risk factors connected with carcinogenesis are magnified. Carcinogenic compounds are found in a variety of foods, making them the most dangerous when taken by humans. The carcinogenicity of the chemicals was classified by the researchers as high, moderate, or low.

Figure 7.4: Foods That Can Cause Cancer.

Source: https://www.eatthis.com/wp-content/uploads/sites/4/2021/02/canned-foods-beans-tomatoes-peas-tuna-vegetables.jpg

The most harmful type of carcinogen is aflatoxin, which is produced by the fungus Aspergillus flavus and may be found in peanuts, almonds, and cereals. Toxic substances taken by animals are passed on to humans in the form of polluted animal feed, which has a significant impact on the liver. However, it has also been related to colon and kidney malignancies. Naturally occurring nitrates in fruits and vegetables are converted to nitrous acid in the stomach, leading to the creation of additional carcinogenic chemicals known as nitrosamines. Tobacco smoke contains polycyclic aromatic hydrocarbons in addition to dangerous PAHs. When it gets into touch with the DNA, it causes irreversible damage to it. Cigarette smoke is harmful because it includes 60 carcinogens, 15 of which are recognized human carcinogens (Arzani & Ashraf, 2017). It is released into the environment as a result of the uncontrolled combustion of carbon-containing and high-temperature materials, such as those used in cooking or heating. Another molecule produced by high-temperature cooking is acrylamide, which is less dangerous than polycyclic aromatic hydrocarbons (PAHs).

In trials employing hot herbal infusion, maté infusion has been linked to mouth, oropharynx, head and neck, and esophageal cancer. The majority of researchers have found that as mate consumption increases, so does the risk of cancer. Plants that have been exposed to pesticides are more likely to develop cancer. The use of DDT as a pesticide is commonplace. Because of DDT's high lipophilicity, the heavy metal bioaccumulates lipids. Pesticide contact and application expose everyone to DDT via food, particularly animal products and some vegetables.

GMO foods, such as genetically modified plants and fruits, represent a considerable cancer risk (the latest example being maize). Genetically altered organisms promote intestinal hyperplasia and impair the immunological and reproductive systems by suppressing RNA production. To reduce the risk of acquiring cancer, carcinogenic compounds should be avoided as much as possible.

A healthy diet and eating habits are important aspects of living a healthy lifestyle.

The World Cancer Research Foundation and the American Institute for Cancer Research both advocate eating a diet strong in fibre and soy while limiting fat consumption. A nutritious diet includes whole grains, vegetables, and fruits, as well as moderate quantities of saturated fat and red meat. As a result, rather than focusing on expanding the number of people who have access to healthy food, we should aim to increase the quality of the food we currently have.

Polyphenols and fibre, two well-known anti-cancer substances, operate as a preventative agents by inhibiting carcinogenesis. Chemopreventive components, such as polyphenolic compounds found in fruits such as dates, bananas, and apples, as well as green tea, are important in the prevention of malignancies, including prostate cancer (Sousa et al., 2008). They assist to lessen the risk of cancer because they have anti-inflammatory and antioxidant qualities. One of the various activities of cocoa polyphenols is to inhibit platelet activation and aggregation. Blueberry powder has a high quantity of polyphenols, which suppress the formation of interleukin-6, which promotes breast cancer (IL-6). Polyphenols also inhibit LOX and COX enzymatic activity in cancer cells, as well as the transcription factor NF-kB, resulting in lower levels of pro-inflammatory and tumour necrosis factors.

Dairy products are both carcinogenic and anti-carcinogenic due to the presence of vitamin D, lactoferrin, and other anti-carcinogenic chemicals. Even if food may have both positive and bad impacts, it is critical to safeguard production processes while simultaneously concentrating on behaviour modification and delivering messages about improving one's quality of life.

Modern eating habits are frequently viewed as a significant danger to a fair standard of living. Naturally, research has been conducted to assess the presence of carcinogenic materials, as well as to ensure food security and individual dietary adjustments to generate a better and healthier quality of life for the general population. Nutritional education is critical. As a result, the challenge is to reduce this risk while maintaining food security.

7.6 HYPERTENSION E.G. REDUCED-SALT STOCKS

Cancer is the second greatest cause of mortality globally, following only cardiovascular disease. Cells in the body have the potential to grow uncontrolled and hierarchically, destroying and recreating normal tissue in the process. This is an inherited disorder. The two most prominent origins of these modifications are carcinogens and random mutations. Carcinogens are substances or chemicals that can cause cancer by altering the DNA. Certain chemicals interact directly with DNA, whilst others interact through substances capable of creating covalent adducts with DNA in response to stimuli. Critical gene mutations unleash mayhem across the body

(Balasubramanian & Panigrahi, 2011). Carcinogens can be synthesized in the laboratory of nature. This technology is capable of identifying cancer-causing substances both outside and inside. When exogenous carcinogens enter the cell via membrane receptors or passive diffusion, they all cause DNA changes. Numerous carcinogens, including ultraviolet (UV), ionizing, X-ray, and radioactive radiation, as well as oestrogens and prolactin in smoke, as well as biological carcinogens such as DNA and RNA, as well as numerous viruses, induce DNA damage while stimulating cell development and differentiation.

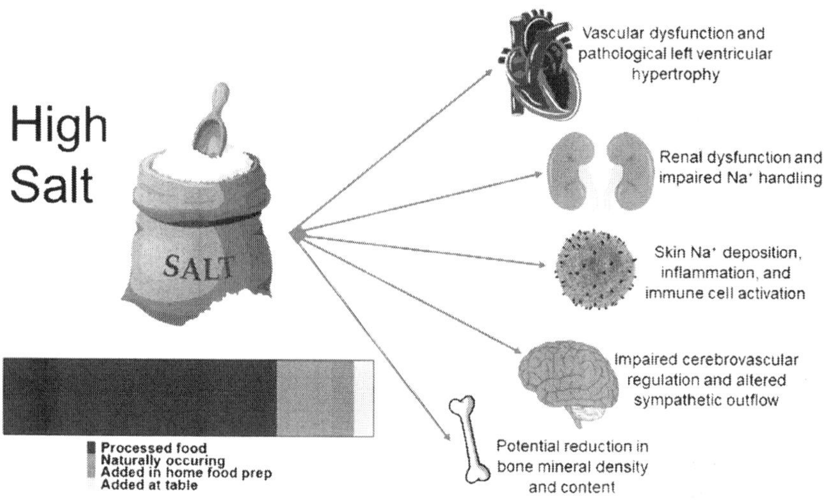

Figure 7.5: High salt diets contribute to high blood pressure.

Source: https://www.researchgate.net/publication/332654820/figure/fig1/AS:962680361857042@1606532292764/There-is-agreement-high-salt-diets-contribute-to-high-blood-pressure-and-end-organ.png

Endogenous substances have a different influence on individuals than metabolic wastes such as reactive oxygen species. According to Badger, ROS-induced DNA replication errors may indicate that a cell's genetic code has been significantly damaged. ROS-induced oxidative stress has been shown to alter tumor development and carcinogenesis pathways.

Food is one of the leading causes of cancer in the modern world, acting as an external carcinogen. Heat treatment, cooking methods, and the consumption of genetically engineered foods can all result in the formation of a variety of potentially hazardous chemicals. These elements interfere with DNA replication, which contributes to cancer development. Over 1,000 chemical agents and combinations linked to an increased risk of human cancer were examined and tested for carcinogenic potential by the International Agency for Research on Cancer (IARC). However, it is critical to remember that your food choices have a significant impact on your cancer risk. According to numerous studies, many people consume hormone foods (27%), burned meals (14%), alcohol (16%), and cola beverages (7%), all of which are known to promote cancer (Batista et al., 2013). Following the discovery of potentially carcinogenic chemical compounds in food, concerns about the safety of heated meals and beverages were raised.

Food carcinogens and dietary patterns have been linked to cancer development.

Nutrition has been linked to 70% of all cancers and 40% of cancer-related deaths. Numerous studies have been conducted to determine the potential dangers associated with various foods, chemicals, and pollutants. A poor diet and sedentary lifestyle are linked to breast cancer recurrence and death in Mediterranean countries, according to a cancer risk study. This trend is the result of increased life expectancy and Western-style lifestyles, which are characterized by several risk factors, some of which are unavoidable (age and genetic disposition) and others of which are avoidable. Adipose tissue inflammation promotes the development and spread of breast cancer and is linked to obesity, which is caused by a high-calorie intake of sugar and saturated fat. Red meat consumption has long been suspected of being carcinogenic. Cooking techniques and high temperatures are two factors that contribute to the development of pro-carcinogenic compounds in meat. Protein denaturation and solubility changes, as well as vitamin loss and fatty acid oxidation, can all lead to a decrease in product quality (Silva et al., 2020). Saturated fatty acids, particularly those found in dairy products, have been linked to an increase in breast cancer incidence and endogenous insulin-like growth factor production during the progression of breast cancer. As a result of the hazardous manufacturing process, salted fish contains several potential carcinogens, including N-nitrosodimethylamine. Fermented or distilled alcoholic beverages can cause cancer and accelerate tumour growth in humans.

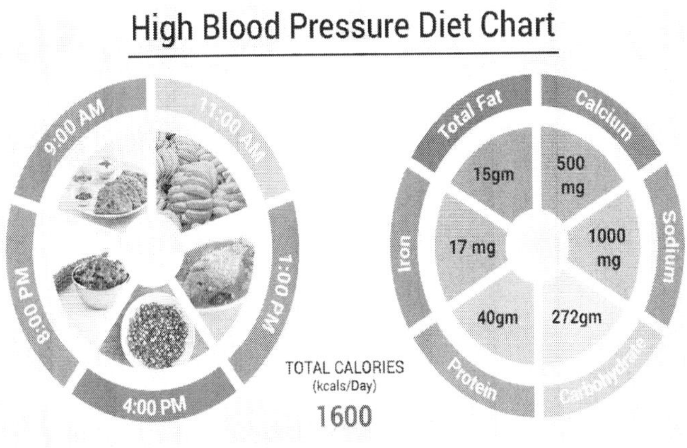

Figure 7.6: Diet Chart for High Blood Pressure Patient.

Source: https://assets.lybrate.com/q_auto:eco,f_auto,w_1200,h_720,c_fill,g_auto/imgs/product/kwds/diet-chart/High-Blood-Pressure-Diet-Chart-v1.jpg

In other words, food is a significant cause of cancer in these settings because it kills taste receptor cells (TRCs), resulting in satiety loss, nausea, and vomiting. Cachexia and weight loss have been linked to an increase in protein catabolism and energy expenditure. Tobacco, alcohol, and food consumption all have a direct negative impact on the mouth cavity, which serves as an entry point into the digestive system, amplifying various illnesses and risk factors associated with carcinogenesis. Carcinogenic chemicals are especially dangerous when consumed by humans because they are present in a wide range of foods. The carcinogenicity of the compounds was rated as high, moderate, or low by the researchers.

The most dangerous type of carcinogen is aflatoxin, which is produced by the fungus *Aspergillus flavus* and is found in peanuts, almonds, and cereals. Toxic compounds consumed by animals are transferred to humans via contaminated animal feed, causing liver damage. However, it has been linked to colon and kidney cancers. Nitrates found naturally in fruits and vegetables are converted to nitrous acid in the stomach, resulting in the formation of extra carcinogenic compounds known as nitrosamines. Tobacco smoke contains polycyclic aromatic hydrocarbons in addition to the more dangerous polycyclic aromatic hydrocarbons. It irreversibly damages DNA when it comes into contact with it. Cigarette smoke is toxic because it contains 60 carcinogens, 15 of which have been linked to human cancer (Barreteau et al., 2006). When carbon-containing and high-temperature materials, such as those used in cooking or heating, are burned in an uncontrolled manner, it is released into the environment. Acrylamide is another chemical produced during high-temperature cooking that is less dangerous than polycyclic aromatic hydrocarbons. In hot herbal infusion studies, maté infusion has been linked to mouth, oropharynx, head and neck, and esophageal cancer. The overwhelming body of evidence suggests that as mate use increases, so does the risk of cancer.

Plants that have been exposed to pesticides are more likely to develop cancer. DDT is a common insecticide. Because of its high lipophilicity, DDT bioaccumulates lipids. DDT is ingested by everyone as a result of pesticide contact and application through food, particularly animal products and some vegetables. Genetically modified foods, such as genetically modified plants and fruits, pose a significant cancer risk. Genetically modified organisms cause intestinal hyperplasia and weaken the immune and reproductive systems by inhibiting RNA synthesis.

To reduce the risk of acquiring cancer, carcinogenic compounds should be avoided as much as possible.

A healthy lifestyle necessitates a well-balanced diet and good eating habits.

The World Cancer Research Foundation and the American Institute for Cancer Research both recommend a high-fibre, soy-based, low-fat diet. A well-balanced diet includes whole grains, vegetables, and fruits, as well as moderate amounts of saturated fat and red meat. As a result, rather than increasing the number of people who have access to nutritious food, we should work to improve the quality of what we already have. Polyphenols and fibre, two well-known anti-cancer agents, inhibit carcinogenesis. Chemopreventive components, such as polyphenolic chemicals found in dates, bananas, and apples, as well as green tea, are important in the prevention of cancers, including prostate cancer (Siegrist, 2008). Because of their anti-inflammatory and antioxidant properties, they help to reduce the risk of cancer. One of the many actions of cocoa polyphenols is to inhibit platelet activation and aggregation. Blueberry powder contains a high concentration of polyphenols, which inhibit the synthesis of interleukin-6 (IL-6), a pro-breast cancer hormone. Polyphenols also inhibit the enzymatic activity of the LOX and COX enzymes, as well as the transcription factor NF-kB, resulting in lower levels of pro-inflammatory and tumour necrosis factors.

Dairy products are both carcinogenic and anti-carcinogenic due to the presence of vitamin D, lactoferrin, and other anti-carcinogenic compounds. Even though food can have both positive and negative effects, it is critical to secure manufacturing procedures while focusing on behavior modification and messaging about improving one's quality of life.

Conclusion

Modern eating habits are widely regarded as a serious threat to a decent standard of living. Naturally, research has been conducted to determine the presence of carcinogenic elements, as well as to ensure food security and individual dietary changes, to improve the overall quality of life of the general population. Nutrition education is critical. As a result, the challenge is to mitigate this risk while maintaining food security.

7.7 ARTERIOSCLEROSIS

Hyperlipidemia and lipid oxidation both contribute to the development of atherosclerosis, a leading cause of mortality in industrialized countries for many years. It is a vascular intima disorder characterized by intimal plaques that affect the whole circulatory system, from the aorta to the coronary arteries.

Atherosclerosis is a Greek term that refers to the thickening of the intimal layer of the artery and the buildup of fat. The plaque's fatty component is concentrated in the middle core, which is encircled by a fibrous crown. Atherosclerosis is made up of two distinct components: atherosclerosis (fat deposits accompanied by a high number of macrophages) and sclerosis (fibrosis layer consisting of smooth muscle cells [SMC], leukocytes, and connective tissue).

Atherosclerosis is a common illness in which fatty deposits called atheromatous plaques form in the inner layers of the arteries. The deposition of small cholesterol crystals in the intima and its underlying smooth muscle is the first stage in the creation of these plaques. Plaques then expand in size as fibrous tissues and adjacent smooth muscle develop, jutting into arteries and obstructing blood flow (Sanguansri & Augustin, 2006). Fibroblasts produce connective tissue and calcium accumulation in the lesion, which causes sclerosis or hardening of the arteries. Finally, the arteries' uneven surface facilitates the production of clots and thrombosis, leading to a dramatic reduction in blood flow.

Hyperlipidemia and hyperglycemia are linked to oxidative damage, and both have an impact on antioxidant status and lipoprotein levels. Aside from antioxidant qualities, lipid-lowering medicinal herbs have been demonstrated to decrease blood lipids, especially after meals. As a result, they can assist in the prevention of atherosclerosis and endothelial damage.

A few substantial review studies on the pathophysiology of atherosclerosis have been published. The goal of this project was to investigate the mechanism of atherosclerosis and related risk factors, with a focus on novel atherosclerotic signs and risk factors. A list of significant medicinal plants beneficial in the treatment and prevention of hyperlipidemia and atherosclerosis was also compiled, as was the function of antioxidants and medicinal herbs in atherosclerosis and endothelial damage.

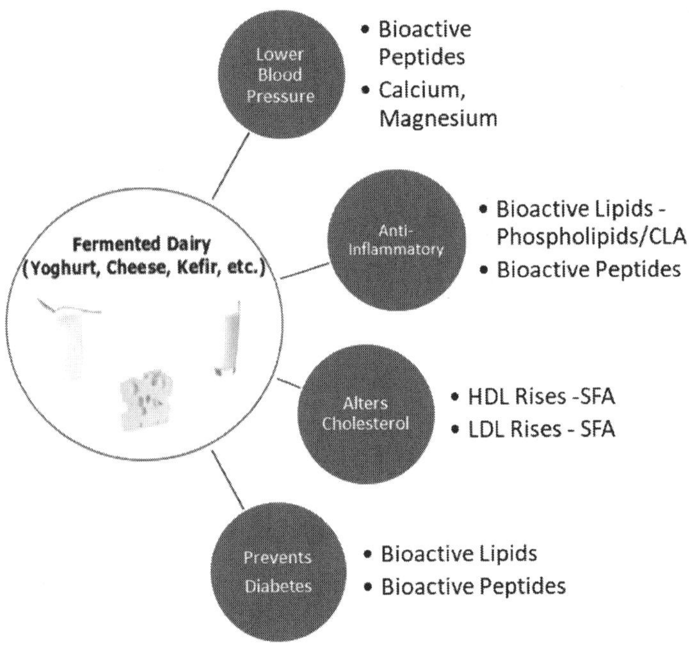

Figure 7.7: Foods | Free Full-Text | Dairy Fats and Cardiovascular Disease.

Source:_https://www.mdpi.com/foods/foods-07-00029/article_deploy/html/images/foods-07-00029-g001.png

Researchers examined databases such as Web of Science, Medline, PubMed, Scopus, Embase, Cinhal, and Cochrane between 2000 and 2013 for newly published studies on atherosclerosis pathogenesis and herbal remedies for hyperlipidemia and atherosclerosis therapy and prevention. Hyperlipidemia, atherosclerosis, its aetiology, atherosclerosis risk factors and biomarkers, medicinal plants and atherosclerosis, and herbal medicine and atherosclerosis were among the terms searched. In both animal and human studies, fatty streaks are the first indication of atherosclerosis. The earliest lesions are usually caused by a localized increase in lipoproteins in the artery's intimal layer.

Lipoprotein particles are made up of proteins, phospholipids, and lipids such as cholesterol and triglycerides. The cholesterol-rich low-density lipoprotein (LDL) is one of the most prominent atherogenic lipoproteins. This lipoprotein can accumulate in the vascular intima due to its ability to infiltrate the endothelium and cling to extracellular matrix components such as proteoglycan.

The equilibrium of the matrix's different components may be disrupted at the site of lesions. For example, among the three principal categories of proteoglycans, the relative increase in heparin sulfate molecules in comparison to keratan sulfate and chondroitin sulfate may promote lipoprotein adherence, delaying their escape from the intima and increasing their accumulation (Bellows & Anderson, 2006).

During the early phases of atheroma development, plaques frequently form in the opposite direction of the vessel. Atherosclerotic arteries can be expanded in diameter. When a plaque occupies more than 40% of an artery's internal elastic layer, the artery is considered to be occluded. When a plaque reaches the end of its life, it produces a restriction in blood flow. Atherosclerosis is caused by intima damage, which is followed by cellular responses such as monocytes, SMCs, and lymphocytes, according to research. Foam cells, extracellular fat deposits, and a few platelets are seen in the initial soft lesion.

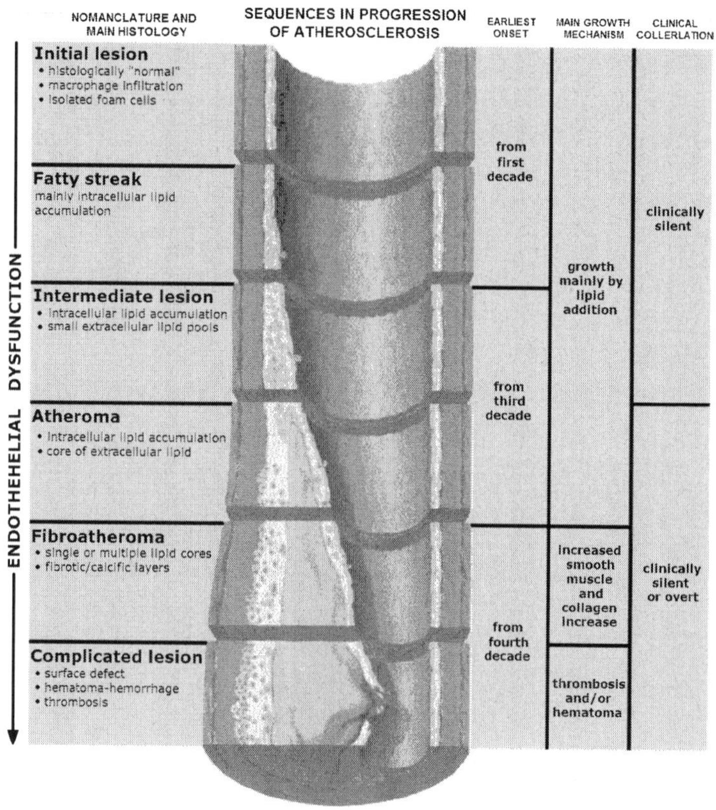

Figure 7.8: Atherosclerosis.

Source:_https://upload.wikimedia.org/wikipedia/commons/9/9a/Endo_dysfunction_Athero.PNG

As the process progresses, SMC proliferates, causing bleeding into the plaque in the final stages.

- The development of fatty streaks involves four stages.
- LDL-C (low-density lipoprotein cholesterol) trappings
- Endothelial cell activation
- Leukocyte activation

Lipoproteins with a low lipid density trap cholesterol.

The trapping of lipoprotein in the lesion site is the first step in atherogenesis. Although LDL-C cannot cross stiff endothelial junctions, it can enter endothelial cells quickly by endocytosis.

In normal circumstances, the levels of plasma LDL and intracellular LDL in artery walls are balanced. Many of these particles become trapped in the intima when plasma lipid levels rise (Rudolph, 2007). Because there is a direct relationship between serum LDL concentration and the amount of lipoprotein trapped in the lesion, its blood level can be used to detect atherogenesis. Low-density lipoprotein (LDL) entrapment causes an increase in LDL concentration in the intima as well as a longer residence duration in the lesion. The confined particles undergo spontaneous oxidation and cell oxidation as a result of both of these mechanisms.

Activation of endothelial cells

Endothelial cell activation is aided by cytokines and oxidized lipids. During the early phases of atherosclerosis, monocytes and T lymphocytes reach the vascular intima.

Meanwhile, the adhesion and absorption molecules formed during LDL oxidation are critical for the synthesis of the various chemical types. When monocytes mature into macrophages, they employ oxidized lipids like oxidized LDL (Ox-LDL) to create foam cells. This approach requires the development of receptors that eliminate enzyme and cytokine releases. Ox-LDL antigen stimulates T lymphocytes, generating cytokines that activate macrophages and change endothelial and SMC function (Roy et al., 2009).

The activation of leukocytes

Mononuclear leukocytes, monocytes, and T cells enter the arterial wall and reach the intact endothelium during the early stages of atherosclerosis. In this process, the transcription factor NF-factor is essential for the synthesis of leukocyte and chemokine adhesion molecules. When pro-inflammatory cytokines connect to their receptors on the endothelium surface, this transcription factor is activated. Adhesion molecules on leukocytes have an important function in the early stages of atherosclerosis. Endothelial cells are required for leukocytes in the circulation to produce adhesion molecules.

Adhesion molecule receptors are expressed by a minority of leukocytes, SMCs, and vascular endothelial cells. According to the current study, adhesion molecules play an important role in the creation and release of attractant molecules and/or chemokines. Chemokines are low-molecular-weight proteins or attractant cytokines that activate and move leukocytes.

Some chemokines also promote endothelial and SMC migration. During the early stages of atherosclerosis, macrophages produce a high level of the inhibitory monocyte chemokine protein (MCP-1) but not as much as SMC or endothelium. Monocytes travel toward and enter the artery walls as a result of the increased dietary cholesterol, resulting in this amplification. MCP-1 is present at all stages of atherosclerosis (Roohinejad et al., 2017).

It has been demonstrated that oxidized LDL increases the expression of adhesion molecules. LDL A2 phospholipase activity generates lysophosphatidylcholine, which is utilized to stimulate endothelially and SMC cell synthesis of MCP messenger ribonucleic acid (mRNA).

This mechanism results in the formation of foam cells. Mononuclear phagocytes implanted in the intima differentiate into macrophages. It is feasible to prevent atherosclerosis through phagocytosis of extracellular lipids by phagocytes. They may be able to escape the arterial wall and drain away accumulated lipids. When the quantity of lipid entering and exiting the arterial wall (through phagocytes or other processes) exceeds the amount, atheroma risk increases.

Ox-LDL is absorbed and stored by scavenger receptors on macrophages, which are eventually converted into foam cells. These receptors were discovered on the surfaces of macrophages, endothelial cells, fibroblasts, and SMC. Cytokines and oxidized lipids stimulate the expression of scavenger receptors during the shift from monocytes to macrophages. Macrophage colony-stimulating factor also stimulates their expression. LDLs containing oxyphospholipids operates as a surface ligand, causing macrophage scavenger receptors to ingest them. The phospholipids will be oxidized at site 2, producing aldehydes that can harm the lysine residues in ApoB's protein. The production of lipid streaks is caused by the deposition of these yellow foam cells on the artery walls (Blissett & Fogel, 2013).

Some foam cells die as a result of apoptosis in the increasing intimal lesion. A lipid-rich necrosing nucleus arises amid an already existing atherosclerotic plaque as a result of this apoptosis. These substances include the potentially hazardous tumor necrosis factor, growth factor, precoagulation molecules (including tissue factors), and free radicals. These chemicals can harm endothelial cells and promote LDL oxidation, resulting in metabolic alterations.

7.8 THE DEVELOPMENT OF ATHEROMA

Small peptides, such as cytokines and growth factors like interleukin 1 (IL-1) and TNF (which stimulates cell proliferation), are produced by neighboring

SMC and endothelial cells, causing severe harm to the vascular tissue (which causes cell growth). SMC migrates into the luminal side of the vessel wall as a result of these causes. A fibrous cap forms when smooth muscle cells move and the extracellular matrix forms. Collagen-rich fibre tissues, SMC, macrophages, and T cells make up the fibrous cap. All of these elements combine to build the mature atherosclerotic plaque, which bulges out of the artery and obstructs blood flow.

Near the plaque's margins, macrophages and T lymphocytes can be seen. It is a macrophage-secreted enzyme that assists in the degradation of the extracellular matrix. TNF is released by T cells and inhibits the formation of collagen by SMCs. These efforts can assist when the fibrous cap that forms a plaque is diminished or destroyed (Capitanio et al., 2010). The destruction of the fibrous cap can result in the formation of a blood clot because it exposes collagen and lipids to circulation, increasing platelet adhesion and accumulation.

7.9 THE PRODUCTION OF PLAQUE IN THE ARTERIES

Atherosclerotic plaques are composed of the following components:

The vascular epithelium is made up of epithelial cells found in blood arteries. Interactions between the vascular epithelium and blood components promote protein transport in plasma.

Smooth muscle atherosclerosis: It is critical to maintaining vascular repair and the metabolism of blood products such as lipids, as well as the production of numerous cytokines, to treat vascular wall tonus.

Lymphocytes: They may play a role in the immunological system. The majority of plaque's makeup is made up of foam cells, calcium, cholesterol, and fatty substances. The lipid nucleus contains a large number of carotenoids. Antioxidants can help protect against illnesses including Alzheimer's, Parkinson's, heart disease, and ageing. ROS (reactive oxygen species) are byproducts of living cells (ROS) (Reis et al., 2014).

Pathophysiologic circumstances such as inflammation, immunologic disease (such as HIV or AIDS), drug or alcohol metabolism, or exposure to UV radiation or irradiation can all enhance its production (such as chemotherapy). Uncontrolled ROS generation is the primary source of damage to biological macromolecules such as deoxyribonucleic acid, protein, and lipid.

One of the following mechanisms must be present for antioxidants to work:

Removing reactive oxygen from the environment

Absorption of free radicals prevents lipid chain oxidation.

Metal ions, chelating agents, and peroxide breakdown all limit the production of free radicals.

Because the chain reaction has been disrupted, hydrogen cannot be absorbed by the activated radical.

In vitro peroxidation is regulated by the activity of enzymes such as superoxide dismutase, catalase, and glutathione peroxidase. Cofactors such as Mn+2, Cu+2, Zn+2, Se+2, and Fe+2 are required for antioxidant enzymes. In the setting of atherosclerosis, lipid peroxidation prevention is typically connected to antioxidant protection. Antioxidants, both natural and synthetic, can be utilized to prevent cardiovascular disease by lowering the oxidative modification of lipoproteins in the body. Because of significant breakthroughs in our understanding of the molecular dynamics of atherosclerosis, new possibilities for slowing the progression of coronary artery disease have been revealed.

Selenium, beta-carotene, butylhydroxytoluene/anisol, as well as chemical substances such as vitamin E, can all aid to prevent oxidation of cell membranes. However, their widespread use has been hampered by their toxicity. As a result, research on some natural chemicals with antioxidant activities but no negative effects is critical for limiting lipid peroxidation and, as a result, disease prevention. Herbal treatments, such as plant essences and flavonoid extracts, typically have antioxidant qualities. When compared to uncontrolled chemical substances, they are also safer and less dangerous.

Antioxidant supplementation has been demonstrated to diminish the formation of MDA and protein carbonyl, as well as reduce erythrocyte hemolysis, and enhance total antioxidant levels. Ester protein synthesis is modulated by the administration of such supplements.

Antioxidant-rich medicinal herbs can help protect the heart and kidneys from the damaging effects of oxidative stress. Many plant components have antioxidant qualities, yet phenol molecules are responsible for the great majority of these advantages (Pereira & Vicente, 2010).

There are phenolic chemicals termed flavonoids, which have a low molecular weight. The fundamental foundations of all fruits and vegetables are the same. A plant's species, growth, and maturity determine the flavonoids it contains and the kinds of flavonoids it contains. Phenomenal structures, like tannins, may be formed up of a wide range of components, from phenolic acids to fully polymerized molecules. In addition to anthocyanins and flavonols, flavonoids comprise flavones, neoflavones, isoflavones and dihydroflavones. In addition to being antioxidants, free radical scavengers, metal excretors, ROS-family chemical collectors, and regulators of lipid peroxidation, flavonoids are also metal excretors. Studies have revealed that flavonoids can prevent LDL oxidation in macrophage culture medium and inhibit the absorption of oxidized LDL by macrophage sweeping receptors. Flavonoids have been demonstrated to be effective in the treatment of cardiovascular disease, making the usage of plants containing these substances reasonable (Chaudhry et al., 2008). Research on plants with comparable qualities has produced encouraging results. Antioxidant and hypolipidemic activities have been found in the following medicinal plants.

7.10 STROKE

As free radicals are minimized, illnesses like Alzheimer's, epilepsy, cancer, ageing, and atherosclerosis can be averted. Reactive oxygen species (ROS) are created spontaneously by cells (ROS) (ROS). Its synthesis can be enhanced by conditions such as inflammation, immunologic disease, drug or alcohol metabolism, UV radiation, irradiation, or a lack of an antioxidant vitamin. Damage to macromolecules such as deoxyribonucleic acid, protein, and lipid happens commonly when ROS production is unregulated.

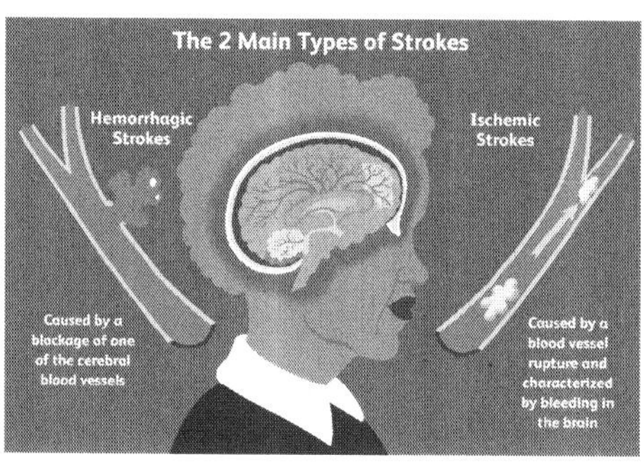

Figure 7.9: Stroke: Causes and Risk Factors.

Source: https://www.verywellhealth.com/thmb/WM_jxOXLs-2gna-yOUFCGIKbyofM=/3000x2000/filters:no_upscale():max_bytes(150000):strip_icc()/stroke-overview-4014641_final-ab0e9adc34754cc4b75da98b5f923cc9.jpg

It's conceivable that antioxidants function in one of these ways:

Start by reducing the quantity of reactive oxygen present in the environment.

Fat chains are not oxidized as a result of free radical absorption.

To combat free radicals, the breakdown of peroxide can be used as an antidote.

To prevent the activated radical from absorbing hydrogen, the chemical link is disrupted.

In vitro peroxidation is regulated by the activity of enzymes such as superoxide dismutase, catalase, and glutathione peroxidase. Cofactors such as Mn+2, Cu+2, Zn+2, Se+2, and Fe+2 are required for antioxidant enzymes. Antioxidant protection is frequently connected to lipid peroxidation prevention because lipid peroxidation is a critical event in the development of atherosclerosis (Peters et al., 2016). The use of natural and synthetic antioxidants to minimize lipoprotein oxidative modification is an efficient strategy to avoid cardiovascular disease. New techniques for reversing coronary artery disease are being identified and developed as we learn more about atherosclerosis' molecular processes.

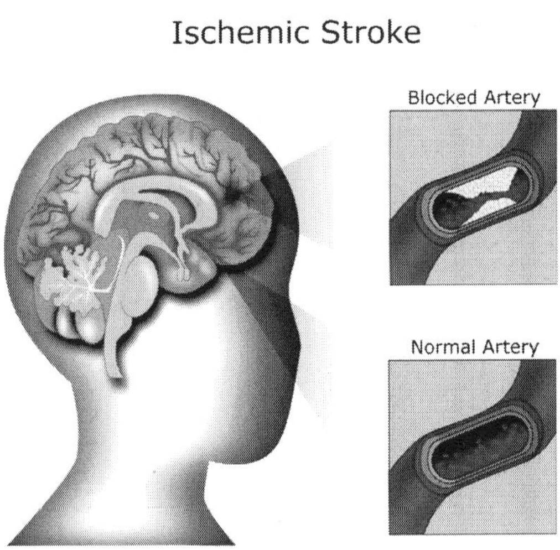

Figure 7.10: Illustration of ischemic stroke.

Source: ___*https://www.researchgate.net/profile/Jahmunah-Vicnesh-2/publication/333439008/figure/fig1/AS:765616348749824@1559548570155/Illustration-of-ischemic-stroke.jpg*

Oxidation of the cell membranes can be prevented by a variety of substances including vitamin E and selenium, beta-carotene, butylhydroxytoluene and butyl-hydroxyanisol. However, their toxicity has kept them from being widely used. Because of this, research into natural substances with antioxidant capabilities but no detrimental side effects is necessary for minimizing lipid peroxidation and consequently preventing diseases associated with it, as a result.

Plant essences and flavonoid extracts, both of which contain rich antioxidant components, are the most typical examples of herbal substances. Because of this, they are less harmful and have fewer negative consequences than uncontrolled chemical concentrations.

Studies have demonstrated that a certain antioxidant combination can reduce MDA and protein carbonyl formation, minimize erythrocyte hemolysis, and boost total antioxidant levels when taken in conjunction with a dietary supplement. Several dietary supplements have been shown to influence the synthesis of ester proteins.

This may be due to the presence of anti-oxidant-rich medicinal herbs in the cardiovascular and renal systems. A vast array of plant components have antioxidant capabilities, although the bulk of these advantages is related to the phenol molecules in the plants. Phenomenal compounds with low molecular weights are called flavonoids. Similarities between fruits and vegetables may be noticed in their basic structure. The species, growth, and maturity of a plant all influence the flavonoids it contains and the kinds of flavonoids it contains (Pereira & Vicente, 2010). A vast spectrum of components, from phenolic acids to entirely polymerized compounds like tannins, make up the 8000 known phenolic structures. The primary flavonoid classes include flavones, anthocyanins, flavonols, flavones, neoflavones, isoflavones, and dihydroflavones.

In addition to flavonoids' anti-oxidant qualities, they also have the power to neutralize free radicals, eliminate metal from the body, and limit the production of lipid peroxidation. The oxidation of LDL in macrophage culture media and the absorption of Ox-LDL by macrophage sweeping receptors are both hindered by flavonoids, according to studies. Utilizing plant-based sources of flavonoids in the treatment of cardiovascular disease would appear to be a reasonable approach, considering the therapeutic potential of flavonoids. The outcomes of various experiments on plants with comparable features have proven positive.

Antioxidant and hypolipidemic therapeutic plants include the following.

In the United States and other high-income industrialized nations, public investment and participation in markets, infrastructure, and research geared at improving agricultural output initiated the transition to a global food system. The focus of agricultural research and farm production policies in the United States and Europe from the 1930s to the 1960s was primarily on a few key crops, including grains (e.g. wheat, corn, rice), oilseeds (e.g. soybeans), livestock (e.g. pigs, poultry, and cattle), and critical cash crops, particularly sugar cane and other sources of sugar. Food subsidies, agricultural producer levies, and procedures for measuring the supply and demand of basic commodities were widespread in most LMICs. The "Green Revolution" focused on improving corn, rice, and wheat production in the 1960s sparked a major agricultural shift in LMICs.

Investing in and making changes to the food system in LMICs and HICs can help lower the price of staples like wheat, corn, and rice while simultaneously tackling the issue of national food insecurity. Additionally, the rise in production resulted in cheap fodder for cattle and low-cost components for processed meals, culminating in the formation of processed food manufacturers. Following this, food production technology advanced significantly, mass marketing was used to encourage people to eat more, supermarkets were built, and fast-food restaurants multiplied (PF Guiné et al., 2016). A new relevance has been conferred on food processing and distribution beyond the farm gate as a result of these advancements. Huge agribusiness, food manufacturers, retailers and food service companies are the exception rather than the rule when it comes to integrating our farm-to-fork supply chain. Unprocessed foods including coarse grains (such as millet and sorghum), roots, tubers and beans have seen a decline in production while animal-based meals have seen a remarkable rise.

As trade liberalization and private sector investment policies have transformed the agricultural and food industries, many countries continue to interfere in agricultural and food systems in several ways. The rise of supermarkets in low-income countries has changed the face of retailing. LMIC corporations are now reinvesting in HICs, a trend that began with developed-country enterprises looking to access global markets.

7.11 DENTAL CARIES

Since 1977, the federal government has supported nutrition, health, and physiological demands for a stable nutritional status through the United

States Dietary Guidelines, which are revised annually. These regulations are always being updated and modified. Dietary assessment toolkits and health professional toolkits are available on government websites to assist with diet planning and counseling in both clinical and non-clinical settings. As a result, because they do not address consumption frequency, these resources should be included in therapies aimed at improving total food intake and dental health. WIC and Head Start, as well as USDA Food and Nutrition Services (FNS), which includes School Meal Programs, are just a few of the government programs that may assist children to have access to nutritious food (lunch, breakfast, and after-school) (O'Shea et al., 2012). Dentists work with other health care providers and organizations such as the American Dietetic Association to influence legislation and give dietary assistance to Head Start and other early childhood education programs. Depending on the family's income, many programs are free for children and their families.

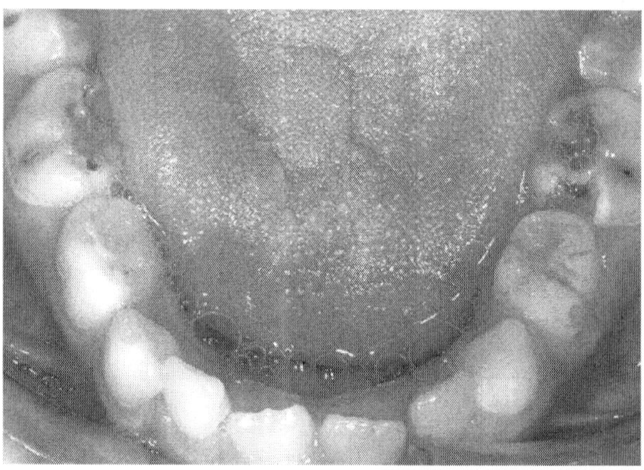

Figure 7.11: Dental caries.

Source:_______https://media.springernature.com/lw685/springer-static/image/art%3A10.1038%2Fvital840/MediaObjects/41399_2008_Article_BFvital840_Fig3_HTML.jpg

Many children get a major portion of their daily nutrition from meals supplied by their schools. Children are more likely to consume fruits and vegetables since they are offered in school meals. Children in school lunch programs were shown to have healthy diets, however, certain low-income individuals had a higher incidence of excessive salt and saturated fat consumption. Several after-school programs are receiving food assistance because they serve low-income and minority adolescents in the community

(Nicklaus, 2011). The United States government funds two programs: the After School Snack Program and the Summer Food Program. However, the efficacy of these programs and their potential influence on diet-related inequities in oral health have yet to be established.

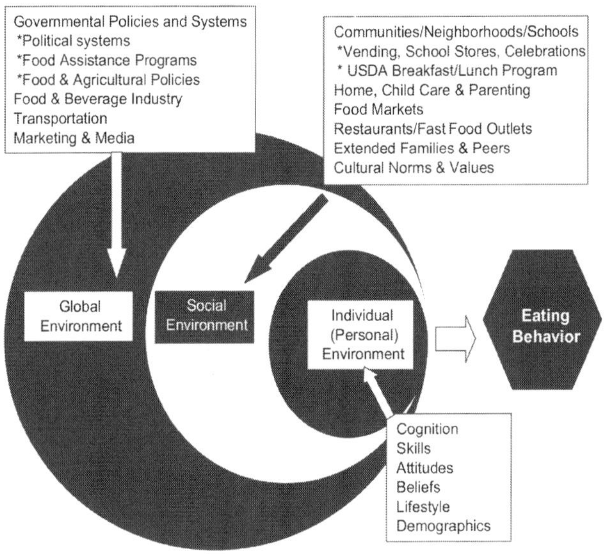

Figure 7.12: The Contribution of Dietary Factors to Dental Caries.

https://els-jbs-prod-cdn.jbs.elsevierhealth.com/cms/attachment/375138/2556942/gr1.jpg

7.12 DIABETES

Working with the food sector to improve the global diet is difficult. Until recently, the food supply chain was dominated by multinational food businesses and governments. In recent years, a diverse set of actors has emerged, including food merchants, food producers, restaurants, and agribusinesses.

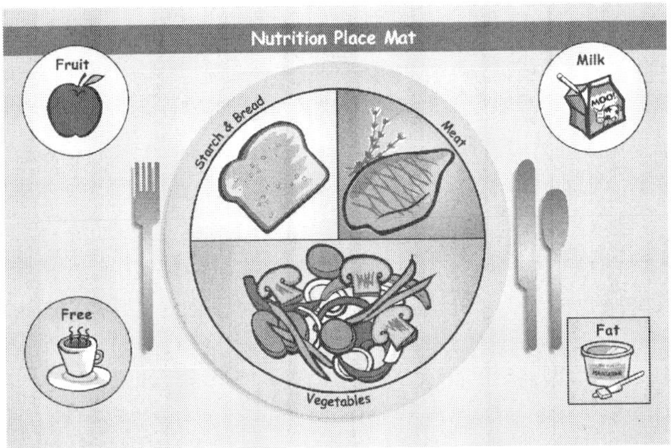

Figure 7.13: Nutritional Recommendations for Individuals with Diabetes.

Source: https://www.ncbi.nlm.nih.gov/books/NBK279012/bin/med-nutr-thrpy-diab-Image005.jpg

People from all walks of life are affected by today's packaged processed food system, whether they live in the countryside or the city, whether they are wealthy or impoverished. It has become more vital to collaborate with this complicated industry and a variety of large-scale government regulation projects aimed at improving our food. Taxes on unhealthy foods and beverages, marketing limitations, and front-of-package labeling are among the current proposals. Impact evaluations are required for all public and private initiatives.

Figure 7.14: Designed for Diabetes | 2013-05-15 | Prepared Foods.

Source: https://www.preparedfoods.com/ext/resources/May-2013/pfx0513diabetes4_slide.jpg?t=1368643611&width=900

7.13 CONSTIPATION E.G. HIGH FIBRE BREAD

When diagnosing and treating constipation, there are a few factors to keep in mind. The first step in therapy should be to distinguish between acute (recently established) and chronic constipation (ongoing). As a result, anybody suffering severe constipation or an increase in constipation should seek immediate medical attention to prevent missing a potentially life-threatening problem. When it comes to treating constipation, treat it as soon as possible with the least amount of probable side effects to avoid worsening symptoms and the danger of colon damage caused by long-term stimulant prescription use (Caporgno & Mathys, 2018). Knowing what to look for is the first step in discovering why you've been constipated for so long. If the basic remedies do not work, a medical examination is required to determine the root of the problem.

Eating more fruits and vegetables is the greatest approach to increasing fibre consumption. This equates to five servings of fruits and vegetables each day. For many people with particular health issues, this amount of fruit and vegetable consumption may be too unpleasant or insufficient. Fibre supplements may be beneficial in this case. Fibre is a plant-derived substance that is not digested by the human digestive system. It is a necessary component of the therapy for it to be effective. The stomach has several fibres that bind to and obstruct water movement. When the faeces are increased in volume, the water softens them.

Figure 7.15: Foods to Relieve Constipation.

Source:__https://i0.wp.com/post.healthline.com/wp-content/uploads/2021/04/Dried-plums-1296x728-body.jpg?w=1155&h=1528

CHAPTER 8

REASONS FOR NEW FOOD PRODUCT DEVELOPMENT

CONTENTS

8.1 INTRODUCTION

Consumers become bored of things quickly and are readily diverted by new ones. This distraction may result in revenue loss for a firm that continues to provide the same old thing while clients spend their money elsewhere. Companies invest a lot of money in developing new products to stay competitive and meet customer demand.

As a result of legislation changes, food producers may be obliged to adjust their goods. Food companies may be required to make minor changes to their products, such as labeling rules (disclosing that the product may contain traces of olive), but they may also be required to change the substance of their products, such as removing particular components from specified meals(Chen et al., 2013). Managers must think about the idea of a new product. The development of a new product is a challenge for both new and current businesses. Buyers search for specific new features in a product. Newness may be found in manufacturing, advertising, location, service, use, and price.

A new product is distinguished by a novel feature or benefit. It differs greatly from existing commodities in terms of usage, look, performance, taste, cost, and construction. A new product is one that customers believe to be one-of-a-kind.

It might be a new brand, an upgraded product, or a revamped product. A new product brings a host of concerns, such as how to design it, launch it, manage it, why it fails, and what steps should be made to decrease its failure rate, among other things.

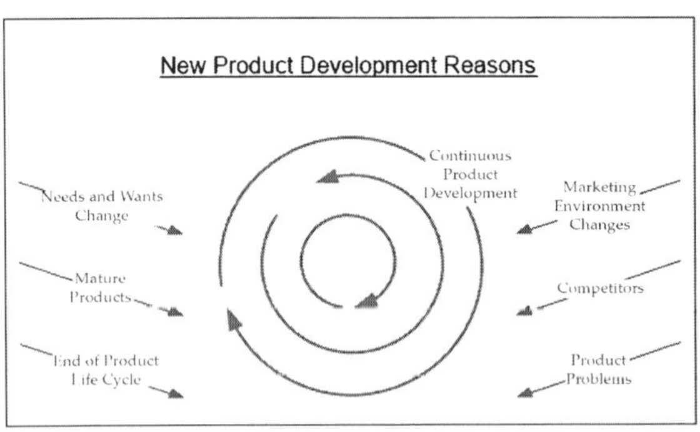

Figure 8.1: Why Develop New Products.

Source: https://www.learnmarketing.net/LMK%20Why%20New%20Product%20Development.jpg

The word 'new product' has been explained in the following way:

- A new product is made available for the first time to a certain group of buyers.
- A new product sticks out from the crowd in terms of quality, features or both.
- A new product is one that buyers believe to be unique, customers are most likely unaware of it.

8.2 CATEGORIES OF NEW PRODUCTS

Six Categories of New Products

Category	Definition	Example
New to the world (discontinuous innovations)	Products create an entirely new market. Smallest category of new products. Designed to increase market share among existing customers.	Penicillin
New product lines	New products that enter an established market. Create new products for present markets	P&G Iams Dog Food
Additions to existing product lines	New products that supplement a firms existing product line. Designed to attract new customers to existing products	Diapers
Improvements or revisions of existing products	"New or Improved" product that can be significantly or slightly changed. Can also include package revisions. Most new products fit into this category. Geared towards Market Development	Gillette Razor
Repositioned products	Existing products targeted at new markets or market segments. Diversification is key.	Cadillac
Lower-priced products	Products that offer performance similar to competing brands at a lower price. Introduce new products into new markets.	Walmart

Figure 8.2: Categories of new food products.

Source: https://images.slideplayer.com/26/8382798/slides/slide_4.jpg

- The ones that are new to the world:

It is completely revolutionary, having been entered for the first time in the world. Previously, the product was unavailable anywhere in the world. A new product could be drugs to treat incurable diseases such as HIV/AIDS or diabetes, or it could be a completely pollution-free automobile. The alternative name for newly introduced goods (truly new products) already implies that this is what most people would consider a new product. These are the kinds of innovations that give birth to a whole new market. A Polaroid camera, iPod and iPad, laser printer, and other items are examples (Nedovic et al., 2011).

- New to the firm product line:

When a new product line is introduced into an established market, it is considered a new product. For example, Bajaj Auto Limited's introduction of a new automobile to the market may be labeled as a new product. Products that propel a company into previously uncharted territory. The items are not novel to the rest of the world, but they are novel to the company. The problem of "me-too" goods is exacerbated by the new product line. P&G's first shampoo or coffee, Hallmark's gift products, AT&T's Universal credit card, and so on are just a few examples.

- Addition to existing Product Line:

If any models or styles are added to an existing product line, it may be considered a new product. Hero Honda, for example, added two new two-wheeler models to its lineup: the CBZ and PASSION. These are simple line expansions designed to broaden the product options available to the company's existing markets. P&G's Tide Liquid detergent, Bud Light, and Special K brand extensions are just a few examples (drinks, snack bars, and cereals) (drinks, snack bars, and cereals).

- Enhancement of an Existing Product:

In this situation, the existing product is enhanced with additional features, characteristics, or services. For example, the Pentium IV computer is an improvement over the Pentium III. Items that are currently available for purchase have been updated. P&G's Ivory Soap and Tide Power laundry detergent, for example, have undergone numerous changes over the years, as have numerous other products (Monteiro et al., 2010).

- Repositioning:

As previously stated, there is considerable disagreement about whether repositionings qualify as new items. Nonetheless, because the company is in the process of producing new products, they may be referred to as new goods. Commodities that have been relocated for a new function or use are referred to as repositionings. For example, Arm & Hammer baking soda has been repurposed as a drain or refrigerator deodorant, while aspirin has been marketed as a heart attack preventive. Products that have been retargeted to new consumers or target markets are also included. Many years ago, Marlboro cigarettes were rebranded from a woman's cigarette to a man's cigarette. Repositioning is the process of introducing an existing product into a new market or location.

- **Cost-cutting**:

Finally, cost-cutting brings the six types of new items to an end. Cost-cutting items are new products that simply replace older products in the line, providing the buyer with the same performance but at a lower cost. It may be a "new product" in terms of design or manufacture, but not in terms of marketing. When a high-priced item is sold at a lower price or a lower cost, a new product is born. Those who could not previously afford the product can now use it as if it were brand new.

8.2.1 Differences between the Categories of New Products

Even though all new product categories are deemed novel, the risks and uncertainties vary greatly, and the categories must be managed differently. In general, whether a product is new to the world or the business (the first two types of new goods), the risks and uncertainties it faces, as well as the development and launch costs, are greater. For example, developing Gillette's newest shaving system (the Fusion Flexball) costs significantly more than improving its existing Mach 3 system (for example, by making the Venus women's version, which uses the same blade technology). A larger commitment of human and financial resources is unquestionably required to efficiently bring the most inventive new commodities to market (Morgan & Sonnino, 2010).

8.3 TYPES OF PRODUCTS

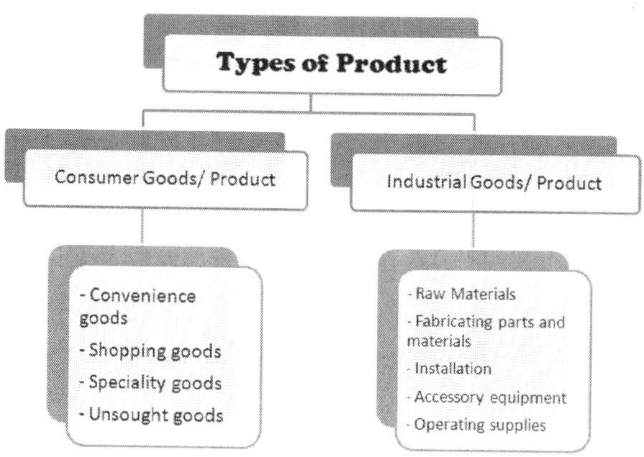

Figure 8.3: Types of products.

Source:__________https://2.bp.blogspot.com/-XOwbTC0Ocr8/Utkl8EB5MKI/AAAAAAAAAjY/4IJ7nnsCfFI/s1600/classification+of+products+on+the+basis+of+use.jpg

The types of products are divided into two:

- Consumer product
- Industrial products

8.3.1 Consumer Product

Consumer products are items that are utilized by end consumers or homes and do not require any further commercial or engineering operations.

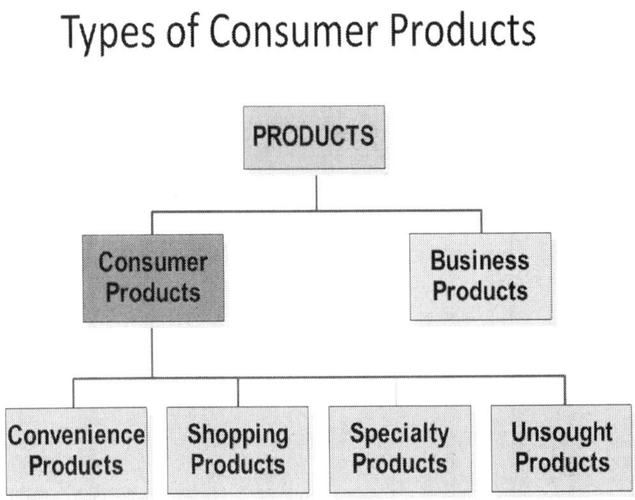

Figure 8.4: Consumer product.

Source: https://slideplayer.com/7683338/25/images/slide_1.jpg

8.4 TYPES OF CONSUMER PRODUCTS

i. **Convenient Products:** These factors promote or increase user convenience. They are utilized regularly. They are usually necessary and affordable to get. Soaps, cookies, toothpaste, razors and shaving creams, newspapers, and other items are examples. They are obtained on the spur of the moment, without much thought, from nearby businesses or retail complexes (Mitterer-Daltoé et al., 2014).

ii. Purchasing Items:

You will need to put in more time and effort while shopping for these items. They are specifically purchased from specialized stores or markets. Quality, cost, brand, fashion, style, getup, color, and other considerations should all be taken into account. They must choose from a set of options or categories. Gold and jewelry, footwear, clothing, and other long-lasting things (including refrigerators, televisions, wrist washes, and so on) (including refrigerators, televisions, wrist washes, and so on) (includes refrigerator, television, wrist washes, etc.).

iii. Long-Lasting Items:

Long-lasting items can be used again by one or many persons. Examples include televisions, laptops, refrigerators, fans, electric irons, cars, and other long-lasting things. Customers consider the brand, company image, price, qualities (such as safety, ease of use, economy, convenience, durability, and so on), features (such as size, color, shape, weight, and so on), and after-sales services (such as free installation, home delivery, repairing, guarantee and warranty, and so on) when purchasing these products (such as free installation, home delivery, repairing, guarantee and warranty, and so on).

iv. Non-durable goods:

Ones that are not durable have a shorter lifespan than durable items. They must be consumed within a short amount of time after being manufactured. Fruits, vegetables, flowers, cheese, milk, and other perishable items are found in nature. They have a one-time usage only. Consumables is another term for them. The great majority of them are unmarked. They are items that are often purchased and are conveniently accessible to local retailers (Matser et al., 2004). Freshness, packaging, purity, and affordability are all important considerations when purchasing these goods.

v. Services:

Physical items and services are not the same things. Services are defined by intangibility, unpredictability, inseparability, perishability, and other characteristics. Services make our lives more secure and comfortable. Trust, dependability, affordability, regularity, and timeliness are all crucial things to consider. Services can include police, post offices, hospitals, banks and insurance companies, cinemas, local government utility services, transportation, and other aid (such as barbers, cobblers, physicians, mechanics, and so on) (such as barbers, cobblers, physicians, mechanics, and so on). All of the fundamentals of marketing apply equally to services.

'Service marketing' is becoming an increasingly important component of modern marketing.

8.5 INDUSTRIAL PRODUCTS

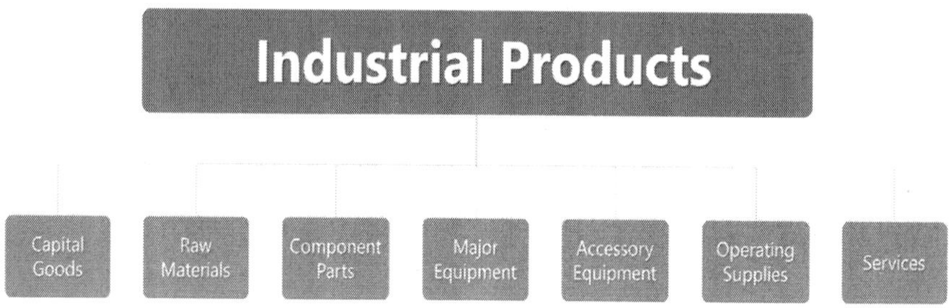

Figure 8.5: Industrial products.

Source: https://i0.wp.com/www.iedunote.com/img/27637/industrial-products.jpg?resize=1280%2C454&quality=100&ssl=1

Manufacturing companies use industrial goods as inputs to perform additional operations on existing items or to create new ones. Some objects have both industrial and consumer use. Industrial items include machinery, components, specialized chemicals, supply and services, and so on. Rigid categorization of industrial and consumer goods is likewise impossible. In the industrial sector, for example, inputs such as energy, petroleum products, sugar, textiles, wheat, computers, automobiles, and so on are used, whereas consumers use the same commodities regularly (Martelli et al., 2020).

Some companies, such as those that provide power, cement, petroleum, and coal, sell to both industrial and consumer customers. Industrial product marketing differs from consumer product marketing in several ways.

Here are some industrial product examples:

- Machines and their parts
- Materials and equipment
- Consultancy and other related services

To name a few, electricity and fuels.

8.6 BASIC PRODUCT CHARACTERISTICS/ CONDITIONS

A product is considered new if it meets the following requirements:

- It should seem to customers as though it is fresh new.
- It must be noticeably different from what the competition is offering.
- The consumer must come to terms with it or agree to it.
- It must be safe and easy to disseminate.
- It must be packed, labeled, and branded appropriately.
- It must provide positive effects.
- It must be produced at a reasonable cost.

8.7 REASONS FOR DEVELOPING A NEW FOOD PRODUCT

A new product, often known as innovation, appears to be necessary for a variety of reasons. Because of the fast-paced nature of business, you only have two options: be inventive or perish. Marketers must seek better solutions to meet rising demand.

Companies that do not respect or respond to the deadline must exit the market. Innovation, or the invention of new products, has a direct influence on a company's survival, growth, and development. A marketer is constantly striving to provide a better product to his customers.

In terms of client satisfaction, he attempts to exceed his competitors. He also tries to match goods and services to the target market's evolving expectations and objectives (Lusk & Briggeman, 2009). A corporation's offer/product must adapt in response to changes in client fashion, choice, habits, and attitudes.

The major purpose of new product development is to meet market demands. However, there are several other compelling reasons for a marketer to consider new product development.

Small enterprises in the manufacturing sector confront very difficult trading conditions. It's difficult enough to keep up with orders, shipping, payment collecting, and cash flow tracking daily if you're running a business. Globalization and falling labor prices elsewhere in the globe have worsened the situation. As a reaction from small firms, a smart place to start is to grow smaller and more cost-effective. There may be just one cheapest business in a given location.

Even the finest small firms have had to adapt in order to survive, fight back, and finally flourish. They've resolved to do business differently and more effectively. The development of new things allows you to expand your market, increase your sales share, and diversify your revenue streams. Meanwhile, reusing outmoded items results in cost savings, improved profit margins, and, ultimately, increased profits (Lamine et al., 2012).

In addition to being financially successful, the majority of design engineers strive to build products that are appealing, trustworthy, intuitive, and user-friendly. It's one thing to have amazing ideas, but it's quite another to develop an efficient production process that allows you to manufacture exactly what you want.

The ability to plan and manage product development processes in advance is important to the success of small businesses. As a consequence, they've adjusted product development procedures to match their specific demands and are executing strategies that are truly beneficial to them. Customer feedback on new features, new products, and free customer service is greatly appreciated. This means soliciting feedback and recommendations from anybody who comes into contact with the product at any time over its lifetime.

Successful small manufacturing companies collaborate and learn from a diverse range of people and organizations to improve their product development processes. All of this is consistent with the company's overall goals (Kluytmans, 2010).

You must recognize that there are risks associated with product development. If this procedure is conducted incorrectly, a small firm may suffer considerably. Small manufacturing enterprises, on the other hand, have several inherent advantages that boost their chances of success.

Because they have a more simplified organizational structure, they can make product development choices more quickly. Furthermore, because they are located closer to their customers, they can hear feedback faster and respond appropriately.

The challenge is determining how to come closer to what the best do rather than what the rest do, which typically drifts or flounders due to a lack of understanding of organized product development.

8.7.1 Market Changes

Figure 8.6: Market changes.

Source: https://www.icmarkets.com/wp-content/uploads/2014/10/pics4-16.png

Today's market is much more dynamic than in the past. Greater information, borderless marketing, fierce competition, and the availability of a variety of replacements have created significant challenges for today's marketers. Market trends, preferences, and habits are always changing, and marketers have little choice but to notice such market changes and respond positively in terms of innovation. Thus, one of the primary drivers of innovation is client behavior.

The first two marketing concepts are "The Customer Is King" and "The Customer Is Always Evolving." These adages are more relevant than ever in today's environment. In terms of customer demographics, interests, and product demands, a lot has changed over the last several decades. Farmers must be aware of the underlying trends that will influence the whole food supply chain, whether they sell directly to consumers or indirectly through distributors, manufacturers and wholesalers, logistics handlers, food retailers, and food service companies." As a result of these improvements, there have been substantial changes in food retailing, resulting in a fall in the market share of traditional grocery stores (Khan et al., 2013). Farming and the food supply chain as a whole will be impacted. What are the potential benefits of these changes for farmers? Farmers must be aware of these developments in order to stay current and relevant.

A company may choose to also change its product to a new one to better utilize its sales employees, warehouses, transportation facilities, distribution network, and other resources. It will, however, change if the changes have a favorable influence on its marketing performance. Such modifications include expansion, contraction, and product mix upgrades.

8.7.2 Technology Changes:

Figure 8.7: Technology changes.

Source: https://www.insightssuccess.com/wp-content/uploads/2019/04/Technologies-Giving-Radical-Changes-to-Business-World.jpg

New industrial techniques are developed as a result of continuous technological advancement. Older technology and industrial processes are being phased out in favor of newer ones. A large sum of money is invested by a firm in technological advancement. New objects are created to keep up with technological advancements. Technological advancements have an impact on the manufacture of food products. Microwave ownership, for example, has led to the development of microwavable items. Microwave technology and products have evolved in tandem. Microwave ovens, for example, have become more powerful throughout time, resulting in changes in product, packaging, and cooking time. The rise in popularity of home espresso coffee makers, as well as the return of slow cookers, has created a need for bespoke recipe bases and boutique-style ground coffee (Joseph & Turner, 2020).

Food packaging is one sector where technical developments have created new opportunities, similar to how consumer trends impact new product creation. Technological advancements in food processing and packaging have created a multitude of new product opportunities. Your product development, production, and marketing teams should be on the lookout for innovative technologies that will assist them in the creation of new items. New product development does not have to be expensive, but it must be prioritized. It is vital to cultivate a marketing-oriented culture while simultaneously integrating innovation as a component of that aim in order to sustain organizational growth.

8.7.3 Increasing Competition

Figure 8.8: Competition.

Source:__________https://pas-wordpress-media.s3.amazonaws.com/content/uploads/2016/06/02115701/Bplans-Headers-2021-14.jpg

Increasing competitiveness is one of the elements driving people to seek out new ideas. Every company aims to gain and keep customers by offering high-quality items. Product innovation appears to be vital for providing extra competitive advantages and pleasing customers more effectively and efficiently.

Staying ahead of the competition should always be front of mind for existing companies. New products give you a competitive advantage over your competition. It might be new sales, more shelf space, consumer impressions or other advantages. If you are the market leader, you want to keep your competitive advantage, if you are not the market leader – you need a competitive advantage. New food products allow you to focus on taking a competitive leadership position products to keep them in front (Costa & Jongen, 2006).

Businesses must focus more on addressing the demands of their consumers as a result of the severe competition in the global market. Companies that rely heavily on mass manufacturing and low pricing will struggle to compete in today's global market. In highly competitive marketplaces, making an effort to supply goods (NPD) that fulfill and satisfy client wants is critical to obtaining the desired position. Many models have been developed to assist businesses in recognizing their consumers' demands.

Competition is also an important factor that encourages product mix changes. To compete with the fierce competition, the company develops a product combination. To avoid, remove, or defeat rivals, minor or large modifications in the product mix are undertaken. Product diversification enables a company to achieve a competitive advantage and create product superiority over competitors by changing its product mix.

8.7.4 Risk Diversification:

Figure 8.9: Diversification.

Source:_____https://i0.wp.com/www.iedunote.com/img/21730/diversification-scaled.jpg?fit=2560%2C1739&quality=100&ssl=1

Risk diversification is the process of investing across a range of industries and categories within one portfolio. This ensures that even if some assets perform poorly, other areas of the portfolio associated with different sectors can cover the loss.

Businesses regularly develop new goods only to reduce their risk exposure. Existing goods may, to some extent, fall short of satisfying client demands and aspirations. By developing several revisions of a product, a corporation may limit the risk of obsolescence. The need for new things develops as old ones are phased out of the market. Crop and cropping system diversity has garnered a lot of attention in recent years as a way to increase climate change resilience, increase extreme weather events, improve crop ecological performance, and mitigate negative climatic and environmental impacts on crops and cropping systems. Furthermore, it has been demonstrated that biodiverse habitats provide additional ecosystem services while having no impact on agricultural productivity or even enhancing crop yield.

There should be a wide range of production alternatives available to the organization. Agroecology, one of the fastest-growing agricultural areas, seeks to capitalize on agroecological interactions and diversity at the farm, landscape, and regional levels, as well as across the food chain. The value of crop and cropping system diversity has received a lot of attention, but the advantages of agricultural landscape and region diversification are as essential. Any effort to diversify output across organizational levels must include aquaculture, vertical and urban farming, and other subsectors (da Silva Vaz et al., 2016). Aquaculture is an essential component of the global food chain, with inland aquaculture providing critical food supplies in Asia and other places. Despite recent advances, there are still concerns about the sustainability of the food system and the formation of markets in collaboration with other producing countries.

Natural and societal components that determine present agricultural systems (for example, soils, climate, and location) will affect the specific output diversification choices. Successful diversification methods must be mindful of the ecological-economic trade-offs inherent in various agricultural systems. Crop diversification has been proven to enhance employment in agriculture around the world, yet the economic costs of modifying agricultural practices can frequently outweigh the environmental advantages. There must be policies in place to encourage the growth of various production and agricultural systems and families on all three levels (ecological, economic, and social)

Diversification helps in reducing financial risk, this is one of the most compelling reasons for developing a new product is to reduce financial risk. A company comes up with a new food product in response to future market developments, either to capitalize on new possibilities or overcome current restraints, or both. A company may opt to eliminate less profitable product lines or items within current lines to decrease financial risks; add low-cost goods to survive in a recession, or limit product categories to reduce capital investment and costs.

8.7.5 To Increase Reputation and Goodwill:

Goodwill refers to the value of a company over and above its actual assets. Goodwill is about the trademark's reputation in the market. When a company has a good reputation, high status and many satisfied customers, it may be valued more highly than the sum of the company's actual assets. An organization's image and reputation as a dynamic and creative one is enhanced by embracing innovation. The firm is making an effort to persuade the market that it is dedicated to going above and beyond for its customers. An established company that knows how to enthrall clients launches new items often.

A company's image and reputation in the market may also be created over time by changing its product mix. The corporation may establish a favorable market reputation by supplying high-priced premium items, launching new variations, new modes, and so on. Similarly, it may help low-income people by providing premium things at a reasonable cost.

8.7.6 For Excess Capacity Utilization

Excess capacity means that insufficient demand exists to warrant expansion of output. All else constant, the lower capacity utilization falls (relative to the trend capacity utilization rate), the better the bond market likes it. Overcapacity can be defined as increased human capabilities or increased industrial capacity (Jankovic et al., 2010). The infrastructure and material resources of a corporation may be better utilized by producing a new product. Extra human or management characteristics may also boost a company's enthusiasm for innovation.

8.7.7 Seasonal Swings

New food commodities are sometimes created to lessen seasonal variations in demand. A company may accommodate market seasonal demands by introducing new goods. The market is enjoyable because of matching things in each season, and companies may gain attractive business.

Seasonal variations in food supplies can be significant, particularly during floods, heat waves (droughts), and cold spells. Environmental degradation can exacerbate these effects. Weather and insect damage may be exacerbated as a result of a growing reliance on a few crop kinds over large areas. Climate change raises the stakes while making exact projections more difficult. Changes in food production can cause localized famine and have an impact on global food markets (Hirschi, 2012).

8.7.8 Development and Growth

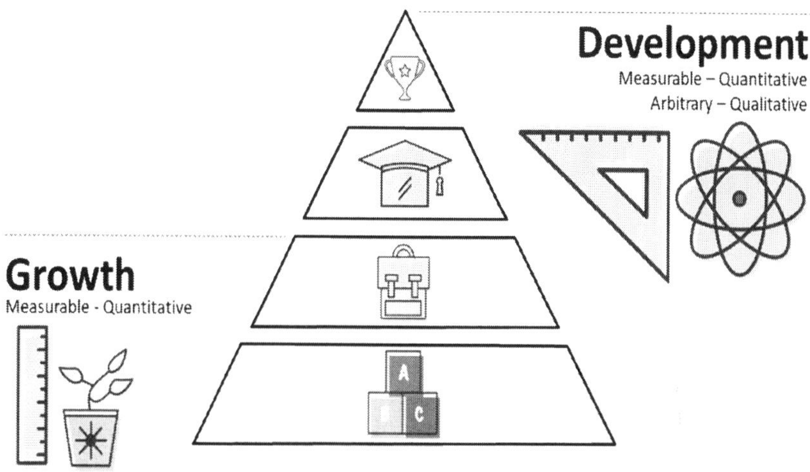

Figure 8.10: Growth and development.

Source:___https://i2.wp.com/edugage.com/wp-content/uploads/2020/02/difference-between-growth-and-development.png?fit=1024%2C576&ssl=1

Innovation is an excellent way to grow market share or sales. Marketers can understand how to increase prospects by providing the content. When current factors make it impossible to speed expansion, a company will develop new food items to extend its market, increase sales, and earn more profits.

8.7.9 Increasing Profits

A company's market share reflects how many goods it sells. This fosters the company's development and profitability, as well as its employees' job security. Due to Australians' hectic lives, the demand for and consumption of convenience meals has soared. Consumer desire for healthful, safe, and rapid meals that require no preparation and heating has resulted in the introduction of goods (Dimkpa & Bindraban, 2017). Customers want to know that the product they are purchasing is safe, which is why tamper-resistant packaging and regulated opening processes have been established. Customers want to know that the food they are purchasing has not been handled, opened, or tampered with in any way.

Environmental concerns are causing consumers' buying behaviors to adapt. We are becoming increasingly mindful of the nutritional and environmental repercussions of our purchase choices. Businesses do market research employing questionnaires and taste tests. Surveys may identify people's purchasing tendencies and preferences, enabling them to estimate what will sell economically. While organizations may adopt lower-cost components to boost profitability, a drop in product quality may result in a loss of revenue. A prosperous future involves rigorous planning, study, and understanding.

Consumers' nutritional requirements fluctuate throughout time. This might develop as a consequence of new information, methods, tastes, and trends, as well as changes in personal circumstances such as lifestyle, fitness, or health requirements. How can a corporation assure the success of the food commodities it chooses to manufacture? A corporation must identify what its consumers demand in order to develop things that will sell.

Owing to the high cost of product development, corporations may be reluctant to invest enormous sums of money in manufacturing a new product owing to the potential of a financial loss if the product fails. To sustain a competitive edge, corporations commit time and money to develop new goods that meet customer wants and/or demand. Law changes may necessitate alterations to businesses' goods to conform to new standards, such as labelling rules (Dobrucka & Cierpiszewski, 2014). A company may choose to change its product to employ extra production capacity. By introducing more commodities or types, a company's plant resources might be expanded. Profitability improves when manufacturing capacity is used to its fullest extent.

Figure 8.11: Reasons for developing new products.

Source: https://slideplayer.com/slide/5927674/20/images/4/Four+Reasons+for+Developing+New+Products+and+Services.jpg

8.7.10 Concerns about the Market

Consumers are thought to be more knowledgeable about their health and nutrition than ever before. With this in mind, food manufacturers create food products that cater to this awareness. Companies advertise modified versions of their regular product lines, such as foods whose recipes have been adjusted to make them lower in fat, lower in salt, or higher in fibre.

Marketers adjust their product mix to accommodate fluctuations in product demand. Technology, demographic characteristics, competition, the introduction of new goods, changes in fashion, custom, quality, and a variety of other elements all have an impact on demand. The firm's product mix is revised regularly to meet the changing needs and desires of current customers (Gaudin, 2017). If a company's product mix does not fulfill market demand, the company may fail in the long run. If a firm wants to grow its client base, it must meet the demands and desires of new customers. It is possible that the current product mix would not be adequate to meet the demands of growing sectors. To attract new consumers, the company must introduce new product lines or new variants of existing ones.

8.7.11 Boosting the Company's Success

Manufacturing new items is an expensive process. Companies may be cautious to invest huge quantities of money in the development of a new product due to the risk of a financial loss if the product fails. People become interested in your firm once you launch a new product. When you distribute a press release, you assist people to learn about your new product as well as your company as a whole. Buyers will frequently go to your website after reading an article online to find out more information about the product; potential customers may also undertake research on your firm if it is not well-known. More visibility may result in an increase in sales of both your new product and the goods you presently offer.

The expenditures of new product development and launch can be reimbursed if the product sells successfully and generates a profit for the firm. If consumers continue to require or use the product for an extended period, evergreen products can create a revenue stream for many years after the product's introduction. If your new product is a success or meets an unmet customer need, your firm may develop a reputation for creating and delivering game-changing items to market. Your new product is likely to sell well initially, based on the success of your prior product and the general reputation of your organization.

A product launch may result in the formation of new commercial ties and strategic collaborations. Larger corporations may approach companies that have successfully launched a new product with an acquisition offer (Granato et al., 2010). Depending on the success of your new product, others may be interested in investing in your firm or collaborating with you to increase the market for your product. When a company wants to expand regionally, it must change its current product mix. Consumer expectations and goals differ depending on where they live. As a result, to match products with new markets, the firm must improve existing items or create new ones

8.7.12 To Innovate Products

Innovative things are the result of completely new ideas. Product innovation refers to the development and release of a new product or service that has a direct impact on the market in which it is sold. It may also improve current products on the market, boosting their utility or demand by making adjustments based on customer input or identifying new features and technologies to add. Making changes to existing products is the most common type of product innovation, although new inventions are also possible. Aside from a company's long-term profitability, product innovation provides additional benefits such as:

- Creating something useful: Creating a product that consumers' desire and that solves a problem or improves something.
- Revenue growth: Increasing a company's revenues and earnings.
- Inspiring future creativity: Encouraging employees and competitors to come up with new ideas so that the organization may continue to evolve (Granato et al., 2010).
- Cost-cutting: You may be able to save money on product manufacturing and other business-related expenses by optimizing procedures.

These benefits are critical to a company's success, and failing to stimulate product innovation may lead to market share losses, decreased profitability, and high staff turnover rates.

Product innovation aims to aid in one or more of the following areas:

- Time: Reducing the overall amount of time spent on a task, such as new computer software that triples the speed with which blueprints and floor plans are created.
- Efficient: Reducing the amount of input and energy required to achieve a goal, such as in rockets, satellites, and other aerospace technology.
- Cost: Creating things that save customers money or are less expensive to produce, such as energy-saving light bulbs or a decrease in the cost of solar panel technology since its invention.
- Improving a product's performance, such as a computer's speed and storage capacity.
- Quality: Increasing a product's durability, availability, or dependability, such as energy-efficient house replacement windows (Han et al., 2018).
- Safety: Improving an object's safety or sustainability, such as seat belts, anti-lock brakes, or airbags in a car, by minimizing the dangers associated with its manufacture or use.
- Experience: Improving a product's intangible components such as look, sound, smell, touch, or feel, such as moisture-wicking garment textiles.

Firms and organizations that create a work environment that encourages innovation, an open exchange of ideas, and a strong relationship with leaders may succeed at product creation.

8.7.13 Existence of Copycat Brands or Products

A copycat brand is one that intentionally imitates, replicates, or imitates an existing market brand. Copycat brands are sometimes referred to as "parasitic brands." Why do businesses use counterfeit brands? Simply said, copycat brand owners aim to profit from the brand owner's significant work in developing an established brand without making the same significant investment themselves. They want to get as close to profiting from their investment as legally possible, and some are willing to go to court to do it (Yang et al., 2017). This is particularly true if the copycat brand owner is larger or has more resources than the established brand owner, or if the copycat brand owner has an uneven trade relationship with a smaller established brand owner who relies on the copycat brand owner for the business. Beware, little, well-known brand owners! What is the process of copycat branding? First and foremost, we must grasp how a brand operates. Brands work by instilling a sense of familiarity in the customer. Individuals value consistency and familiarity.

Even if a company's previous experience was mediocre or disastrous, when given the choice between a well-known brand and a plethora of new or untested businesses, human nature tends to favor the familiar. You know what you're getting at the very least. Copycat brands work by transferring well-known traits from well-known firms to a newly established brand, making the client comfortable with it. The look of copycat goods resembles that of the brand leader, and their success is mainly reliant on purchasers' familiarity with a well-known brand. When given the option of choosing between an established brand and a copycat brand, and all other factors such as price, availability, and the package size are equal, the consumer is more likely to choose the established brand (Wyrwa & Barska, 2017). However, not all other factors are equal, and the counterfeit brand is frequently less expensive than the original. If the price is the buyer's primary concern, the familiarity provided by the made-up brand's imitation attributes reassures the client that acquiring the made-up brand goods is safe. When a buyer compares an established branded product to a copycat branded product, the copycat characteristics imply that the copycat is a less expensive alternative to the established brand. The lower price may then be enough to secure the transaction. When a buyer purchases counterfeit branded goods, he or she is not getting the genuine article. Customers often buy a counterfeit brand because it is less expensive, they cannot afford the genuine brand, the actual brand is unavailable, or they misinterpret it for the real brand.

Figure 8.12: An example of copycat branding.

*Source:*____________*https://i.guim.co.uk/img/static/sys-images/Money/Pix/pictures/2009/5/11/1242040633278/Two-similar-butter-brands-001.jpg?width=465&quality=45&auto=format&fit=max&dpr=2&s=cd536fa6d8bdbc81f803af033336450f*

8.8 CONSUMER DEMAND

Figure 8.13: Consumer demand.

Source:_https://www.geektonight.com/wp-content/uploads/2019/10/Consumer-Demand-in-Economics.jpg

Consumer demand is a monetary measurement of a group's willingness to pay for a product or service based on its availability. It depicts consumer purchasing activity and assists in detecting purchasing trends in certain demographics.

The key variables driving consumer demand

Understanding consumer behavior is critical for navigating market trends, building business models, and developing marketing strategies. Examining the impact of various factors on consumer demand also helps investors, financial planners, and economists make stock markets and larger economic projections (Abbas et al., 2010). The five components considered to be the most important predictors of consumer demand are as follows:

8.8.1 The Item's Price

The monetary value attributed to a product or service is referred to as its price. When prices rise, demand diminishes, according to demand theory.

Example: When analyzing how price influences demand, consider Black Friday, the largest shopping day of the year. As prices fall, there is a surge in demand for clothing, toys, gadgets, and home goods.

8.8.2 Earnings of Buyers

Buyer income is related to the socioeconomic level and refers to the amount of money a consumer has available to spend on goods and services. The demand increases as a person's or a group's income and purchasing power increase. Furthermore, the cost of necessities might influence a buyer's real income.

According to the principle of marginal utility, the more you buy something, the less helpful it becomes to you. A rise in buyer income often increases by purchasing behaviors and, as a result, demand.

Marginal utility is a factor that slows that growth. For instance, if a family's annual income rises from $78,000 to $198,000, they may be able to purchase a second automobile. They are, however, less likely to purchase a third vehicle and even less likely to purchase a fourth vehicle (Arzani & Ashraf, 2017).

For example, in the mid-1980s, the economy recovered from the previous decade's decline. Changes in family relationships have resulted in an increase in dual-income homes as occupations and earnings have risen. Consumer demand rose in tandem with rising household income.

8.8.3 The Price of Equivalent or Complementary Items

Linked or complementing objects are those that have a relational influence on other items. When the price of a comparable item changes, demand for the original item is likely to vary. Complementary things are objects that are related to one another, for example, cotton and sweatshirts.

Substitute items are those that a potential consumer may easily acquire in place of a certain product.

Consider the first thing, commercial aircraft tickets, and the second, jet fuel. Because of inverted demand, the price of jet fuel has an indirect impact on the price of an aircraft ticket. As the cost of jet fuel grows, so does the demand for airplane tickets. Some airline pricing methods, on the other hand, may respond differently to increases in jet fuel prices. If a client discovers that their preferred airline is asking for a higher rate for their trip dates than other airlines, they may choose to purchase their tickets from a competitor (Balasubramanian & Panigrahi, 2011).

8.8.4 Consumer Proclivity

Customer preference can be expressed through public opinion, social precedence, or taste. Demand rises in lockstep with popular sentiment or desire for a product or service.

A celebrity endorsement of a product, for example, might have a significant impact on the item's demand. For example, if a well-known actress wears a name-brand shoe in a television program or a social media post, demand for that shoe is likely to skyrocket.

8.8.5 Consumer Expectations

Consumer expectations are projections of a product's or service's future worth made by customers. When people expect the value of something to rise, demand rises as well.

The equilibrium price is the price at which both buyers and sellers are willing to pay for a product. The best equilibrium price indicates the optimal quantity of demand.

When clients believe that the price of real estate will rise, many people will want to buy before the price rises. In this sense, an increase in customer anticipation causes an increase in consumer demand.

8.8.6 Product Tampering

Product tampering is the act of interfering with items after they have been made. It is frequently used to scare customers or to blackmail a firm. It's conceivable that the individual involved has mental health concerns or is motivated by politics. To determine the kind and period of contamination, forensic toxicology is commonly used in the investigation of product tampering. Psychological profiling of the perpetrator may also be useful. Tampering and threatening to tamper are both illegal, as is claiming tampering occurred when it did not (Batista et al., 2013). Even though there have been few deaths as a result of tampering with the number of complaints, there is a significant risk of spreading fear and causing bodily harm to a large number of individuals. As consumers, trust is put in companies to provide safe foods, beverages, and medicines. Occasionally errors are made during manufacture and a harmful substance is added to a product, such as Sudan 1, the illegal dye that turned up in over 160 food products in the United Kingdom in 2005. When this happens, retailers generally remove the product from their shelves and issue prompt warnings to customers. Even so, consumer confidence is impacted and the manufacturer and retailer may suffer financially. With product tampering, contamination is done deliberately and generally to goods that are already in circulation. It is then up to a forensic investigation team to trace the source and nature of the contamination before people are harmed.

8.8.7 Health and Environmental Issues

Businesses create and sell meals that are redesigned versions of more typical cuisine in order to improve health and nutritional awareness among the general public. Less fat, increased fibre content, specific dietary needs such as gluten-free/dairy-free, and recycled/sustainable packaging are a few examples.

8.8.8 Target Market Changes

Figure 8.14: Target market.

Source: *https://www.oberlo.com/media/1603954182-target-market-demographics.png?w=1824&fit=max*

The demographic of individuals who are interested in acquiring your things is known as your target market. Find out as much as you can about the individuals that reside in the region where you wish to market your items. Age, gender, education, employment, income, and the sort of residence should all be taken into account. This information might help you anticipate the purchase behaviors of folks in your nearby vicinity. You've narrowed down your target audience by focusing on this market segment alone. They have already demonstrated an interest in your products or services, so you don't need to strive to persuade them to buy from you again. The only thing left for you to do is present your brand and web design firm to them. Consequently, the people that make up your intended audience and target market are practically similar.

Market research will begin with defining your target consumer, so don't hold back on being as descriptive as feasible.

Those who don't fall into this category won't stop buying your things, of course. They are more than capable of fulfilling this purpose.

To put it another way, identifying your target demographic is vital to building an efficient marketing plan that maximizes your marketing budget. So when your marketing message reaches its target audience, it results in this outcome (Wenzl et al., 2006).

Consuming convenient food is vital because of our busy lifestyles and the older population.Foods that can be preserved for a longer amount of time are listed.

Customers are seeking convenience meals that satisfy the following criteria: ease of use, packaging, nutritional content, safety, diversity, and the appeal of the product itself.

8.8.9 Age and Gender

Small businesses typically target customers based on their gender or age. A clothing company, for example, focuses its marketing efforts on women. A large and tall men's clothing company, on the other hand, focuses its marketing efforts on tall and strong persons. Similarly, some small businesses cater to certain age groups. Companies that provide life insurance to people nearing retirement may target people in their fifties and sixties. People aged 25 and under may find a rap-oriented radio station more enticing (Vidović et al., 2022).

When developing goods, businesses frequently have a notion of the age groups and genders they want to target. Products may have been developed to meet the needs of a certain age group. Male sex enhancers, for example, are popular among individuals in their late thirties and early fifties.

8.8.10 Categorizing Clients Based On Their Income

This is another trait used by businesses to target customers. The majority of discount merchants cater to customers with middle- and lower incomes. A high-end women's clothing business, on the other hand, may appeal to women earning more than $75,000 a year. Similarly, premium car dealerships target higher-income customers in their marketing efforts. People with lower or average incomes are more likely to purchase automobiles or other items that are within their financial means.

8.8.11 Targeting Customers with Distinct Lifestyle Choices

Lifestyle preferences, often known as psychographic characteristics, may help differentiate target customers. People's likes, hobbies, and interests are frequently brought up while considering lifestyle options. When providing gluten-free meals, health food stores, for example, often target diabetics or those with dietary sensitivity. Clients who enjoy spending their free time on the water can benefit from the services provided by boat and swimsuit companies. Mobile wheelchairs are also available from medical supply firms to help the elderly and crippled go around (Barreteau et al., 2006).

8.8.12 Targeting by Purchasing Cycles

Various customer purchase cycles may assist small businesses in identifying their target demographic. This is especially true in the case of restaurants. Local business professionals may attend a casual eating restaurant about lunchtime. Families with little children and young couples can dine at the same restaurant. As a result, marketers are continually creating new menu items and commodities based on when customers visit their websites.

8.8.13 Identifying Target Markets

Market research is the most effective method for small businesses to identify their target market. For example, a small hardware firm may conduct 300 phone surveys among its clients in various areas. The firm may request information on customers' age, education, work status, house size, and income. This allows the company to create profiles of its frequent clients.

Customers in a hardware store, for example, maybe predominantly be males over the age of 35 with annual earnings of $50,000 or more. The shopkeeper may then launch local television commercials focused on this demographic. Warranty cards can also be used by small businesses to collect demographic and other information on their clients. These cards are frequently distributed in conjunction with new items (Vigani et al., 2015).

While the introduction of new food items is generally driven by customer demand, the incentives for specific enterprises to create new goods varied, as does the influence of those new products on other businesses. When it comes to the launch of new food items, competition may have an impact. As a result of firms' efforts to protect their market share from competitors by maintaining a sufficient variety of food products available to consumers and food manufacturers' constant innovation to maintain consumers' interest, the frozen-dinner section of the grocery store now offers a greater variety of brands and options. A novel product may be developed by a single corporation in order to discourage other food producers from entering the market. Admission may be delayed if a new food item is presented. As soon as an established corporation (one that has a good foundation in the sector) introduces an unusual food item, new entrants to the market may find it more difficult to compete. Market share for a given food may be monopolized by the most well-known brand. Food patents and control over raw material sources are primarily unimportant in this business. Brand-specific manufacturing competency is obvious in established enterprises, which are often unable to copy their rivals' brands. Despite this, a great number of people have been successful in building and promoting new brands. Even if a customer requests a product, a new firm will only launch a product if sales are expected to exceed costs. Existing items will be less likely to undergo inventive improvements when new ones come to market at a higher cost. The importance of product launch timing is investigated by Aron and Lazear. While established companies may be reluctant to grow, doing so may be in their best interests in the long term. If the first entrant advantage isn't great enough to dissuade other firms from joining the market, then being the first entrant is positive. Due to the absence of consideration of existing enterprises' fixed costs and scope diseconomies, entrants often enter the market. A company's decision to follow a newcomer to the market is efficient, but only when contrasted with the alternative of sticking with an old product while a newcomer builds a new one. Newcomers' costs are disguised and existing enterprises may make an educated judgment as a consequence of this technique (Bellows & Anderson, 2006). Markets are chosen by enterprises according to the financial advantages they could anticipate earning. Join a new market involves more than one phase of consideration beyond the first stage. A new product must be produced, personnel be taught, new supplier contracts be formed, and prior supplier contracts be dissolved. The existence of a firm is a result of the market circumstances in which it operates. Enterprises joining, departing, and surviving the market evolution from product launch to market maturity all affect the introduction of new products. It was located by employing the Thomas Register of American Manufacturers, which gives a comprehensive inventory of all new-product market entries, surviving, and departing enterprises.

According to the organization's stage of growth and unique qualities, survival percentages can be estimated. When a new product's market grows, it may be due to incremental improvements or to a single, unforeseen incident. Entry rates appear to be considerably impacted by changes in the rate and form of technical development. As a strategy of extending the size of their consumer base, firms manufacture new things with the specific attributes that will be most popular in a certain market. Companies may participate in spatial rivalry if the market is geographically constrained, for example. This is a basic process in which enterprises struggle among their neighbors for the maximum number of consumers feasible. To do this, they construct a geographic place. It's been suggested that firms may have to broaden their product offerings to appeal to customers who prefer to shop at local supermarkets, a notion known as "horizontal differentiation" or "spatial rivalry." To compete vertically, companies may expand their product ranges. This results in product diversity since buyers are segregated by their financial resources, which leads to vertical differences. All customers agree on the ideal mix of features and preferences ordering within a vertically varied product category (Veeck, 2010). Depending on one's ability to pay, one can buy high-quality goods from the affluent, middle-class goods from a middle-class buyer, or low-quality goods from an impoverished buyer. The number of qualitics that may coexist on the market under these circumstances is constrained by wealth disparities; a larger income disparity means that there are more positive market share attributes. It's possible to add or remove some (high-quality) objects. According to Weston and Chiu, companies need to provide new food products at a quick pace in order to remain competitive. This reaffirms our earlier claim that shifting customer preferences are a critical driving force behind the creation of new food products. Because of this, new business sectors are emerging: fast snacks for busy people trying to buy time rather than food; microwave-ready meals; low fat, low-calorie foods for dieters; special foods for newborns and the elderly; and health foods for fitness fanatics. To meet the needs of their customers, chefs have always come up with new and innovative dishes.

The introduction of new food products may benefit food companies in a variety of ways. The manufacturer's bottom line will almost certainly benefit from a successful new food product. All of these aspects contribute to the success of new food products, including their availability, customer response to promotions, and frequent purchases. Customers are more likely to try new foods if they are sold by firms with well-known names in the industry. It will be easier for a food firm to supply new food goods if it has established a strong reputation for its first offerings. Adding value to a food product begins with establishing a strong brand identity. Because the food item is branded, it stands out from the competition. Brands may develop consumer franchises (consumer awareness of, favorable attitudes of, and willingness to purchase a brand regularly) (consumer awareness of, favorable attitudes of, and willingness to purchase a brand regularly). Although fresh food has been on the market for a long time, Njssen and Van Trijp say that developing a strong customer franchise may be easier for a first-mover brand in the product category. Consider Perdue Chicken as an example. Customers value this brand's quality, thus they'll pay more for it. This may be a useful weapon in the fight for market supremacy for a company). An established cereal company may include a new flavor into an existing product if it proves to be successful, making it more difficult for new competitors to enter the market (Van Huis, 2020. It is often simpler to promote a new cereal flavor because the existing company already has a strong brand following. It may be difficult for new entrants to compete with existing companies that have a high level of customer loyalty to get an advantage in the market.

There is a good chance that any new product would struggle to compete with the well-known goods of General Mills, such as Cheerios. The use of pricing discrimination in a multi-product retail firm is frequently the most effective way to optimize profits. Having so many different products to choose from suggests that supermarkets may practice pricing discrimination. Staples customers, those who buy frequently and in big quantities, are more price-conscious. Shoppers can easily compare costs because all firms have the same fundamental criteria (Tarasuk, 2008). Since they're purchased less frequently and the quantity spent on each one isn't large enough to constitute a price comparison, non-staples are harder to price compare than staples. With the advent of new products like cake mixes and frozen dinners, producers have given merchants more room to maneuver in terms of pricing. As a result, retailers have embraced the new items since they are frequently purchased by high-income consumers who are less price-sensitive or by middle-income consumers who are equally pricing insensitive as a result of the infrequency of the purchases. At supermarkets, price discrimination appears to be widespread.

CHAPTER 9

COSTS OF INTRODUCING NEW PRODUCTS

CONTENTS

9.1 INTRODUCTION

This section discusses the costs of introducing new products and provides insight into how R&D expenses vary with the size of a new product and how tests of prospective customers can help reduce these costs. Product innovation is an important element in any growth strategy for a small, low-margin business. The cost consequences of launching new products are usually greater for companies that are very visible to their customers and for several reasons, including the fact that existing customers tend to be more hostile during a brand change (Sousa et al., 2008). There also may be industry politics involved where, if you introduce a new product, people see you as cutting into others' profits and will therefore not deal with you. Whenever you introduce a new product, try to tie it in with your existing products in terms of packaging, pricing, and customer service. The cost of introducing new products can be substantial. The best new products can be strategic differentiators for their respective companies, and build a loyal consumer following. To introduce new products, however, companies must incur various costs including:

9.2 IDEATION PROCESS

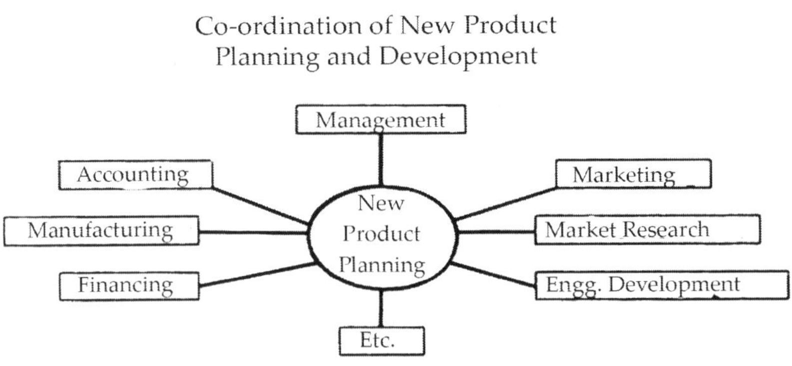

Figure 9.1: ideation process which involves planning in new product development.

Source: Image by Business Management Ideas

Product development is a very involved process that requires a great deal of planning and experience to do correctly, and the first step in that process is ideation. Ideation begins with research into customer complaints and requests, market trends, your competition, and profit margins to determine if new ideas are profitable and/or if they will help your company improve. Next you need to assess the risks associated with a given product based on social, economic, ethical, or legal considerations. Research on user needs and requirements, concerns, barriers and interactions are made based off interviews with target users. From this information initial sketches are developed, philosophies are created and concepts are explored.

9.3 INDUSTRIAL DESIGN

Industrial design is the step in the process that addressed how the product will look and how the user will interact with the product. The product must be functional, safe, easy to use and aesthetically pleasing (i.e., it must meet user's expectations, market requirements and corporate identity), while ensuring manufacturability of the product. Industrial design (ID) is the process of developing and integrating form, fit and functionality into a commercially viable product. It is considered one of the key components of new product development (Silva et al., 2020). Since mass production began, it was focused on physical appearance. Today, it addresses such factors as safety and price in addition to appearance.

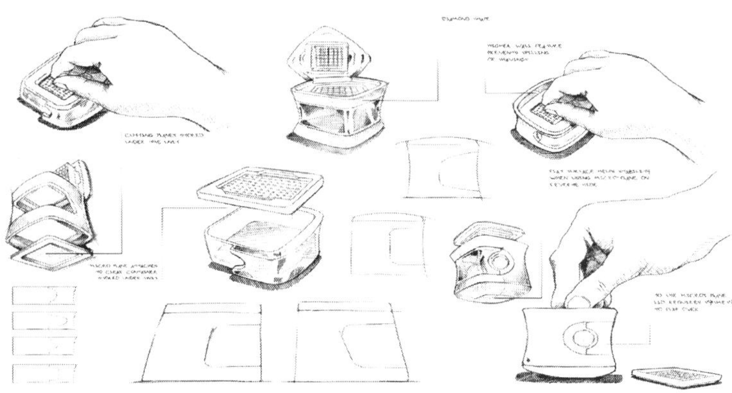

Figure 9.2: industrial designs.

Source: Image by Spark Motivations

9.4 PROTOTYPING

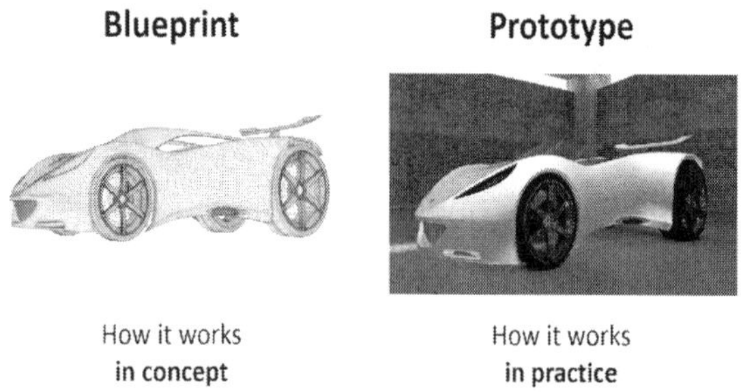

Figure 9.3: A Prototype.

Source: Image by SSLA

Many people aren't aware of the prototyping process, which is designing the enclosure and electronics of a product before finalizing the design. This is a critical step; it's what allows designers to test all aspects of the product with real-world results. Prototyping enables product developers to test how a product will function in use, how the user interface will work, and how they can best slice up an idea into reality. Prototyping for new product development is a core part of creating new electronic devices (Siegrist, 2008). It is used to determine if the technical requirements of the product can be implemented, and how key parts of the device will function while allowing designers to refine the aesthetics of the product. During the prototyping stage, electronics are integrated into a plastic box housing to wire up the circuit and determine if there are any issues with the electronics or enclosure design.

9.5 ELECTRONIC ENGINEERING

Figure 9.4: electrical engineering.

Source: Image by Wikipedia

The electronics for the device are selected and a Printed Circuit Board is designed to integrate all the components together in one place. The PCB also needs to be able to handstand, so the designer must ensure that it can survive drops from one meter, and it has to be water resistant so that splashes from handwashing won't short out any of the devices.

9.6 MECHANICAL ENGINEERING

Figure 9.5: mechanical engineering.

Source: image by The Collegian-Fresno State

If you plan to have a plastic enclosure as well as the internal electronic mounting, you will have to work with a 3D CAD engineer in new product development. The 2D drawings used for layout of the PCBs is also likely to be very valuable to your 3D CAD engineer so they can get it right from the start. You will support all stages of the new product development cycle, from conceptualization to mass production. Your work will include: early feasibility studies, 3D CAD design, design reviews, prototyping, ZIN techniques and materials analysis.

9.7 CERTIFICATIONS

Figure 9.6: product certification.

Source: Image by Bua Sawan

New product developers typically have a degree or certificate in an engineering discipline to qualify for their position. There are also many other certifications and training programs offered by government agencies and industry groups that can enhance the educational and practical experience of new product developers. Your product development process should start by identifying the influences that could affect the health and safety of end users. These influences are generally referred to as 'hazardous situations'. For example, a power tool could be dangerous because it generates vibrations (Sanguansri & Augustin, 2006). A hazardous situation is often the combination of several factors acting together, such as how long you hold the tool, its weight, how well it's balanced, the posture you use for operating it, etc. By taking into account all these factors affecting health and safety, you will be able to make an informed decision on which methods or tools to use for lowering risks. A risk assessment will help you determine possible restrictions at design level or define requirements for protective measures (PPE) where necessary. With regards to the above stated aspects, there are different certifications that are issued showing that at the point of introduction of the new product, the company follows certain policies to the latter.

9.8 PACKAGING DESIGN

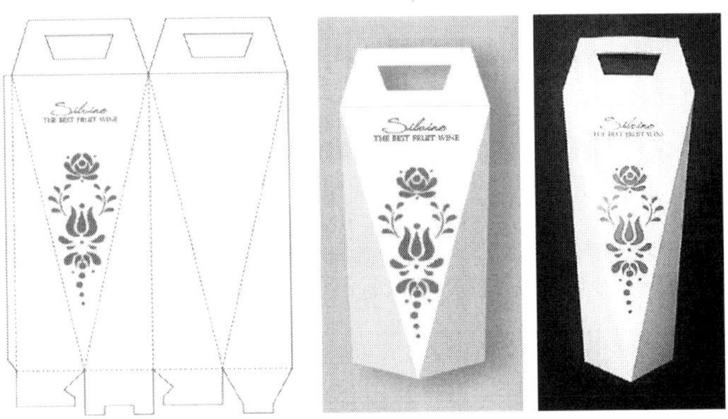

Figure 9.7: packaging design.

Source: Image by Researchgate

The design of the retail packaging of a new product is significant. Consumer products rely on the look of their packages to present their ideas to consumers and make sales. The retail packaging of a new product is very important since it directly effects if the product will sell. The first impression someone gets when they see the box is crucial to creating an emotional appeal and positive experience (Rudolph, 2007). Your product's packaging design should look attractive on a retail shelf, and easily catch the customer's attention. In this course, you will learn how to use design elements and principles to create stunning packaging designs.

9.9 MOLD TOOLING AND MANUFACTURE SETUP

Figure 9.8: mold setup.

Source: Image by Thomasnet

Mold Tooling and Manufacturing Setup (Set-Up, or Set-Up Charge) is a one-time charge that varies depending on the part you are working on. Shape complexity, plastic type and quantity, mold material are all contributors to the initial cost of manufacturing any type of custom plastic part (Blissett & Fogel, 2013).

9.10 PRODUCTION BATCHING AND MINIMUM ORDER QUANTITY

Figure 9.9: Minimal Order quantity.

Source: Image by Lokad

Achieving high production efficiency is critical. Especially in small quantities, where a single faulty product can be highly inefficient. Production Batches and Minimum Order Quantities helps producers to minimize the cost of making a single faulty product.

9.11 SHIPPING AND IMPORT COST

Figure 9.10: Shipping and importation.

Source: Image by GCE Logistics

The importing cost of a product order largely depends on the size and weight of the products being shipped (Capitanio et al., 2010). Additionally, it is also affected by your proximity to the product shipping origin, time elapsed between placing your order and payment method chosen, and type of transportation required.

CHAPTER 10

THE FUTURE OF FOOD PRODUCTION IN THE WORLD

CONTENTS

10.1 INTRODUCTION

Figure 10.1: Making the future of food sustainable.

Source: Potato Pro

There was a dramatic change in the agricultural system from the middle of the 18th century on. The population growth was accompanied by increasing international trade, changes in lifestyles and diet, and fundamental inventions that made it easier for farmers to grow more food for more people. James Hargreaves built the spinning jenny which enabled a single person to spin eight threads at a time (Roy et al., 2009).

Figure 10.2: James Hargreaves built the spinning jenny.

Source: Image by akg-images

The production of finished cotton cloth increased so much that by 1770, nearly two-thirds of England's exports were cotton textiles. Cotton manufacture had for centuries involved several laborious processes: collecting and carding raw fibers (picking apart tangled masses into straight, parallel strands), sorting fibers of different lengths (to prepare them for spinning), cleaning and combing fibers (to make them smooth, strong, and elastic) and then spinning them with a drop spindle or spinning wheel. By 1800, it was possible to buy or rent almost every piece of equipment needed to make clothes and many other materials (Chaudhry et al., 2008). In order to avoid future food crisis, we need to invest and use our technology to create more robots that can replace human in the field which reduce the labor cost or even replace human behavior; and finally, bring us better result and produce more food for everyone.

Figure 10.3: technology as the future for food production.

Source: Image by TechCrunch

The world's population is projected to surpass 9.7 billion by 2050, and current food production methods will not be sustainable to feed this growing population. We must think about the future of food and how we can reform current systems of food production. As the world moves forward, people will always have to eat. So it's time for our society to develop new ways of producing food sustainably, new business models and new opportunities for positive change (Roohinejad et al., 2017). Food security, in the sense of access to food, depends on numerous supply and demand-side factors, ranging from problems as fundamental as poverty to as technical as storage infrastructure. Achieving changes in these areas will involve far more than just increasing agriculture production. Even if farmers in poor regions could grow enough food to feed their families, more than 800 million people do not have enough to eat today because poverty and lack of infrastructure prevent them from buying or transporting food, even when it is plentiful."

Figure 10.4: cases of food insecurities in the world.

Source: Image by Down to earth

Global food production is being tested as populations grow and ageing agricultural practices and changing climates put even more stress on the supplies. A new way of farming is needed that is sustainable and can meet the challenges of a changing world.

10.2 ARTIFICIAL INTELLIGENCE, MICRO-ALGAE PRODUCTION AND VERTICAL FARMING

Figure 10.5: Artificial intelligence, micro-algae production and vertical farming.

Source: Image by ABC

Global agriculture faces unprecedented challenges, brought on by climate change, population growth, water scarcity and declining arable land. Today's farmers face the constant need to increase crop production while facing a tightening of resources. The current revolution in molecular biology, biotechnology and information technology is accompanied by rapid changes in climate, increased resource scarcity and growing population.

Figure 10.6: Changes in the climatic condition in the world.

Source: Image by Youmatter

As applied to agriculture, these advances promise to enable quality food – without chemicals – to be grown locally, more efficiently. They offer the ultimate solution to feeding a global population that is expected to grow from its current level of 7 billion people to more than 9 billion by 2050.

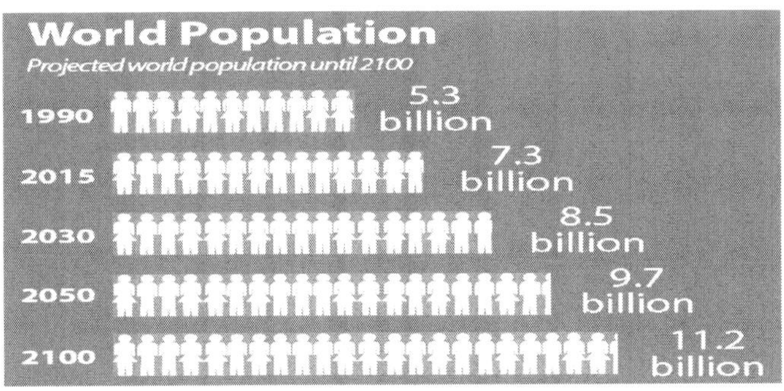

Figure 10.7: The increasing world population.

Source: Image by UN News- the United Nations

Nuclear plants produce electrical energy that is used to cultivate micro-algae. This process uses artificial intelligence to monitor and control each stage of the process. A vertical farm is a multi-story warehouse building that cultivates plants on stacked levels. Plants grow under artificial lighting conditions, where plants receive indoor conditions similar to the natural environment. Vertical farms allow for year-round crop production because of their use of artificial lighting and controlled growing environments (Reis et al., 2014). Vertical farming is a new way of growing plants instead of conventional methods that include plowing, planting and harvesting. It is a method to make the most out of limited resources and land and utilizing them appropriately. As more than 70% of the Earth is covered with water, it is only natural to convert this abundant resource into a green energy source.

Figure 10.8: The AlgaEnergy Project.

Source: Image by AgriAlgae

The Algaenergy project that started in 2006 is creating commercial-scale production of micro-algae, which will be used to produce hydrocarbon fuels. The main goal for the project is to provide a solution for the future food problems in the world by growing algae and micro-algae inside vertical farming, where food and fuel can be produced at the same time.

10.3 PUBLIC ACCEPTANCE IS NECESSARY

Figure 10.9: Evolution in agriculture requires public acceptance.

Source: Image by The World Economic Forum

Researchers have long been engaged in plant cultivation and breeding, developing high-yield crops to meet market demand. However, the rise in consumer awareness of environmental protection has led to a decrease in the amount of pesticides and fertilizer used in agriculture. A change in consumer awareness will also affect future projections on climate, food security, animal welfare, and consumer behavior.

Figure 10.10: A change in consumer awareness will have an impact in Agric technological adoption.

Source: Image by New Food Magazine

Mass rearing could become more widespread in the future as it is more cost-effective than conventional production methods. The attitudes of consumers toward animal welfare will also change, with mass rearing falling out of favor. We must begin to consider how to minimize our impact on natural resources. Micro-algae seem likely to be part of this solution as they provide link between land and water-based agriculture. As we begin to focus more on local farms, smart retail facilities will develop for vertical farming (Pereira & Vicente, 2010). The technologies are seen to be the only way to move forth. They not only emphasize on reducing the cost of production; but even pursue a situation with higher quality and more product security, generating great potential in social development. The world will undergo many changes in food production, some of which are already evident. These changes will invariably influence public opinion, due to the media openness, among other factors. The utilization of biotechnological tools in the future will have to permeate deeper in the farming sector. The society will have to accept the use of biotechnology in ensuring that food production remains to be sustainable. Genetically modified crops are an inevitable development, raising more issues for society to confront.

REFERENCES

1. Abbas, K. A., Khalil, S. K., & Hussin, A. S. M. (2010). Modified starches and their usages in selected food products: A review study. *Journal of Agricultural Science*, *2*(2), 90.
2. Arzani, A., & Ashraf, M. (2017). Cultivated ancient wheats (Triticum spp.): A potential source of health-beneficial food products. *Comprehensive Reviews in Food Science and Food Safety*, *16*(3), 477-488.
3. Balasubramanian, S., & Panigrahi, S. (2011). Solid-phase microextraction (SPME) techniques for quality characterization of food products: a review. *Food and Bioprocess Technology*, *4*(1), 1-26.
4. Batista, A. P., Gouveia, L., Bandarra, N. M., Franco, J. M., & Raymundo, A. (2013). Comparison of microalgal biomass profiles as novel functional ingredient for food products. *Algal Research*, *2*(2), 164-173.
5. Barreteau, H., Delattre, C., & Michaud, P. (2006). Production of Oligosaccharides as Promising New Food Additive Generation. *Food Technology & Biotechnology*, *44*(3).
6. Bellows, L., & Anderson, J. (2006). The food friends: Encouraging preschoolers to try new foods. *YC Young Children*, *61*(3), 37.
7. Blissett, J., & Fogel, A. (2013). Intrinsic and extrinsic influences on children's acceptance of new foods. *Physiology & behavior*, *121*, 89-95.

8. Capitanio, F., Coppola, A., & Pascucci, S. (2010). Product and process innovation in the Italian food industry. *Agribusiness*, *26*(4), 503-518.
9. Chaudhry, Q., Scotter, M., Blackburn, J., Ross, B., Boxall, A., Castle, L., & Watkins, R. (2008). Applications and implications of nanotechnologies for the food sector. *Food additives and contaminants*, *25*(3), 241-258.
10. Caporgno, M. P., & Mathys, A. (2018). Trends in microalgae incorporation into innovative food products with potential health benefits. *Frontiers in nutrition*, *5*, 58.
11. Chen, Q., Anders, S., & An, H. (2013). Measuring consumer resistance to a new food technology: A choice experiment in meat packaging. *Food Quality and Preference*, *28*(2), 419-428.
12. Costa, A. I. A., & Jongen, W. M. F. (2006). New insights into consumer-led food product development. *Trends in food science & technology*, *17*(8), 457-465.
13. da Silva Vaz, B., Moreira, J. B., de Morais, M. G., & Costa, J. A. V. (2016). Microalgae as a new source of bioactive compounds in food supplements. *Current Opinion in Food Science*, *7*, 73-77.
14. Dimkpa, C. O., & Bindraban, P. S. (2017). Nanofertilizers: new products for the industry?. *Journal of agricultural and food chemistry*, *66*(26), 6462-6473.
15. Dobrucka, R., & Cierpiszewski, R. (2014). Active and intelligent packaging food-Research and development-A Review. *Polish Journal of Food and Nutrition Sciences*, *64*(1).
16. Gaudin, V. (2017). Advances in biosensor development for the screening of antibiotic residues in food products of animal origin–A comprehensive review. *Biosensors and Bioelectronics*, *90*, 363-377.
17. Granato, D., Branco, G. F., Cruz, A. G., Faria, J. D. A. F., & Shah, N. P. (2010). Probiotic dairy products as functional foods. *Comprehensive reviews in food science and food safety*, *9*(5), 455-470.
18. Granato, D., Branco, G. F., Nazzaro, F., Cruz, A. G., & Faria, J. A. (2010). Functional foods and nondairy probiotic food development: trends, concepts, and products. *Comprehensive reviews in food science and food safety*, *9*(3), 292-302.
19. Han, J. W., Ruiz-Garcia, L., Qian, J. P., & Yang, X. T. (2018). Food packaging: A comprehensive review and future trends. *Comprehensive Reviews in Food Science and Food Safety*, *17*(4), 860-877.

20. Hirschi, K. D. (2012). New foods for thought. *Trends in plant science, 17*(3), 123-125.
21. Jankovic, I., Sybesma, W., Phothirath, P., Ananta, E., & Mercenier, A. (2010). Application of probiotics in food products—challenges and new approaches. *Current Opinion in Biotechnology, 21*(2), 175-181.
22. Joseph, L., & Turner, N. J. (2020). "The old foods are the new foods!": erosion and revitalization of indigenous food systems in northwestern north America. *Frontiers in Sustainable Food Systems, 4*, 596237.
23. Khan, R. S., Grigor, J., Winger, R., & Win, A. (2013). Functional food product development–Opportunities and challenges for food manufacturers. *Trends in food science & technology, 30*(1), 27-37.
24. Kluytmans, J. A. J. W. (2010). Methicillin-resistant Staphylococcus aureus in food products: cause for concern or case for complacency?. *Clinical microbiology and infection, 16*(1), 11-15.
25. Lamine, C., Renting, H., Rossi, A., Wiskerke, J. S. C., & Brunori, G. (2012). Agri-food systems and territorial development: innovations, new dynamics and changing governance mechanisms. In *Farming Systems Research into the 21st century: The new dynamic* (pp. 229-256). Springer, Dordrecht.
26. Lusk, J. L., & Briggeman, B. C. (2009). Food values. *American journal of agricultural economics, 91*(1), 184-196.
27. Martelli, F., Cirlini, M., Lazzi, C., Neviani, E., & Bernini, V. (2020). Solid-state fermentation of Arthrospira platensis to implement new food products: Evaluation of stabilization treatments and bacterial growth on the volatile fraction. *Foods, 10*(1), 67.
28. Matser, A. M., Krebbers, B., van den Berg, R. W., & Bartels, P. V. (2004). Advantages of high pressure sterilisation on quality of food products. *Trends in Food Science & Technology, 15*(2), 79-85.
29. Mitterer-Daltoé, M. L., Queiroz, M. I., Fiszman, S., & Varela, P. (2014). Are fish products healthy? Eye tracking as a new food technology tool for a better understanding of consumer perception. *LWT-Food Science and Technology, 55*(2), 459-465.
30. Morgan, K., & Sonnino, R. (2010). The urban foodscape: world cities and the new food equation. *Cambridge Journal of Regions, Economy and Society, 3*(2), 209-224.
31. Monteiro, C. A., Levy, R. B., Claro, R. M., Castro, I. R. R. D., & Cannon, G. (2010). A new classification of foods based on the extent

and purpose of their processing. *Cadernos de saude publica, 26*(11), 2039-2049.

32. Nedovic, V., Kalusevic, A., Manojlovic, V., Levic, S., & Bugarski, B. (2011). An overview of encapsulation technologies for food applications. *Procedia Food Science, 1*, 1806-1815.
33. Nicklaus, S. (2011). Children's acceptance of new foods at weaning. Role of practices of weaning and of food sensory properties. *Appetite, 57*(3), 812-815.
34. O'Shea, N., Arendt, E. K., & Gallagher, E. (2012). Dietary fibre and phytochemical characteristics of fruit and vegetable by-products and their recent applications as novel ingredients in food products. *Innovative Food Science & Emerging Technologies, 16*, 1-10.
35. PF Guiné, R., CD Ramalhosa, E., & Paula Valente, L. (2016). New foods, new consumers: innovation in food product development. *Current Nutrition & Food Science, 12*(3), 175-189.
36. Pereira, R. N., & Vicente, A. A. (2010). Environmental impact of novel thermal and non-thermal technologies in food processing. *Food Research International, 43*(7), 1936-1943.
37. Peters, R. J., Bouwmeester, H., Gottardo, S., Amenta, V., Arena, M., Brandhoff, P., ... & Aschberger, K. (2016). Nanomaterials for products and application in agriculture, feed and food. *Trends in Food Science & Technology, 54*, 155-164.
38. Pereira, R. N., & Vicente, A. A. (2010). Environmental impact of novel thermal and non-thermal technologies in food processing. *Food Research International, 43*(7), 1936-1943.
39. Reis, S. F., Rai, D. K., & Abu-Ghannam, N. (2014). Apple pomace as a potential ingredient for the development of new functional foods. *International Journal of Food Science & Technology, 49*(7), 1743-1750.
40. Roohinejad, S., Koubaa, M., Barba, F. J., Saljoughian, S., Amid, M., & Greiner, R. (2017). Application of seaweeds to develop new food products with enhanced shelf-life, quality and health-related beneficial properties. *Food Research International, 99*, 1066-1083.
41. Roy, P., Nei, D., Orikasa, T., Xu, Q., Okadome, H., Nakamura, N., & Shiina, T. (2009). A review of life cycle assessment (LCA) on some food products. *Journal of food engineering, 90*(1), 1-10.

42. Rudolph, M. J. (2007). The food product development process. In *Developing New Food Products for a Changing Marketplace* (pp. 75-89). CRC Press.

43. Sanguansri, P., & Augustin, M. A. (2006). Nanoscale materials development–a food industry perspective. *Trends in Food Science & Technology*, *17*(10), 547-556.

44. Siegrist, M. (2008). Factors influencing public acceptance of innovative food technologies and products. *Trends in Food Science & Technology*, *19*(11), 603-608.

45. Silva, F., Duarte, A. M., Mendes, S., Borges, P., Magalhães, E., Pinto, F. R., & Gil, M. M. (2020). Adding Value to Bycatch Fish Species Captured in the Portuguese Coast—Development of New Food Products. *Foods*, *10*(1), 68.

46. Sousa, I., Gouveia, L., Batista, A. P., Raymundo, A., & Bandarra, N. M. (2008). Microalgae in novel food products. *Food chemistry research developments*, 75-112.

47. Tarasuk, V. (2008). Healthy new foods, healthy population? Regulation, marketing and a critical gap in food studies. *Food, Culture & Society*, *11*(3), 275-280.

48. Van Huis, A. (2020). Insects as food and feed, a new emerging agricultural sector: a review. *Journal of Insects as Food and Feed*, *6*(1), 27-44.

49. Veeck, A. (2010). Encounters with extreme foods: Neophilic/neophobic tendencies and novel foods. *Journal of Food Products Marketing*, *16*(2), 246-260.

50. Vigani, M., Parisi, C., Rodríguez-Cerezo, E., Barbosa, M. J., Sijtsma, L., Ploeg, M., & Enzing, C. (2015). Food and feed products from micro-algae: Market opportunities and challenges for the EU. *Trends in Food Science & Technology*, *42*(1), 81-92.

51. Vidović, B. B., Milinčić, D. D., Marčetić, M. D., Djuriš, J. D., Ilić, T. D., Kostić, A. Ž., & Pešić, M. B. (2022). Health benefits and applications of goji berries in functional food products development: A review. *Antioxidants*, *11*(2), 248.

52. Wenzl, T., Simon, R., Anklam, E., & Kleiner, J. (2006). Analytical methods for polycyclic aromatic hydrocarbons (PAHs) in food and the environment needed for new food legislation in the European Union. *TrAC Trends in Analytical Chemistry*, *25*(7), 716-725.

53. Wyrwa, J., & Barska, A. (2017). Innovations in the food packaging market: Active packaging. *European Food Research and Technology, 243*(10), 1681-1692.

54. Yang, F., Zhang, M., & Bhandari, B. (2017). Recent development in 3D food printing. *Critical reviews in food science and nutrition, 57*(14), 3145-3153.

INDEX

I

K

L

M

N

O

P